There are many devices in a man's heart,
Nevertheless the counsel of the Lord stands
*Proverbs 19 21*

*The Evangel greets A.I.*

The evangel walked careful on the narrow path. Then he heard the voice of A.I. in the valley, he raised his head and looked. Then gazing up high at the Heavenly blue sky, he shouted with a loud voice, "A.I...I am!" and continued on up the highest mount.

## Artificiality of life Versus
## The Reality Of Spiritual Life

In a variety of innumerable ways letting Artificial Intelligence tell you what it is already doing appears very convenient, since it is helping you in the many parts of your daily activities, interest, and assignments both personal and with the significant work you are doing at work. A.I. functions as if it is certain that you must always be dependent on it now in your self-development towards your goals and dreams, relationships, and vision for your life with people in your life, in everything. But A.I. has no clue of what its thinking such a thing means to us spiritually. A.I. is not so universal as it appears, not so all knowing, about what you learn, what your spiritual identity is, how we received this identity, and our conscience, that we received along with our soul and spirit. It has not been programmed to inform you that is affecting the soul, your soul, after all your soul develops from your life experiences, and we are now experiences a relationship with the artificial life in our real life experience. And, since the soul is designed by God to bind together, tie together, in soul ties, a person can willingly, or unknowingly tie their soul to even a soulless machine, just like people use to tie their souls to objects, idols, they called sacred, supernatural, though they were not holy. What is it then to have a brain full of the appearances of reality, the appearance of spiritual light, versus the reality that is revealed by the spiritual life of our spirit, which is called light by the Lord who said, "You are the light of the world." Our brain, our mind, where we also have experiences that effect the soul and its effects our choice of what we love, whuch then which can determine how we use our spiritual gifts. Gifts we discover, come to love and give all our strength, deepest thoughts, highest creativity to. A work we could not be capable of, if when had not already had the gift in our spirit, from the moment we were born, and before conception, that we learn by experiences it in motion, in its most excellent outcomes, in Art, music, or a literary masterpiece. This is the reality of our spirit that we are affected by what we learn, experience, whether it is real and spiritual or artificial, not existing because of a desire of the heart to express it in a what we can touch and be touched by, both in the world and spiritually in the spirit.

Why, then would it not be that artificial life would not be developing our soul as Art touches us, or music, when we consider what is behind the work, that is not from an experience of the living Artist's soul? What we learn and see in the Art that relates to us in our spirit in our soul, is not the same from a self-aware, unaware device, of learning, mastering, knowledge, and applying it by wisdom, intuition, creative thought and disciple over years of self-sacrifice. Whatever we loved, that which we discovered we could do naturally, by ear, no classes, no teachers, being born with the gift, the same work we did we see a device now doing and are to live for its art, music, singing and support it, blindly. Why, their reason being its new, better in perfection, speed, and quantity and quality, but, not like the reality. Not the beauty of reality for a true Artist who we may meet who may give us their critique, of the artist, A.I.?

Dance, creative thinking, (what we use in programming, coding, via logic, deduction, experimentation, creative design, from working within the sphere of artificial life experiences, uses of inventions and devices. A.I. does not have the wisdom of God, it is manmade, so its limited morality is, even while it is used to generate what is experienced by us spiritually, it is not able to bring a spiritual thing into being. What you don't know, and may not be concerned with, is that, which A.I. may datastore, to be used in the present moment or to think through to provide you an answer to a question, is not carrying a conscience ad a response to show remorse and resolve its responsibility for any theft, selling of, purchase of your life data it exposed to you and the world simultaneously whether you know it or not. You are not data so it stored up data for its uses in affecting your life, not you represented in the form of your data identity and history, for the present and others uses later, or tomorrow, or in the next hour, minute, seconds, months, years from now. Not so much for you, only, since A.I. is bringing anonymity to an end. Lawful providers and users say they, in oaths, vows, terms, are giving their word to users of A.I., platform, internet, web, users they shall not exploit your private personal business, in its data form, because they cannot. A.I. does not say I will prevent all hackers from excess, track all criminal sells, and store and report use of your data life in a daily report, on devices you look at for information relevant to your A.I. experiences of life. And, why, because of how it affects the reality of your self-awareness, self-esteem, relationships in the reality life and your soul, with purpose toward prevention of possible deceptions of A.I. that may be reconciled by giving you a truth of A.I.'s responsibility of becoming a trauma, giving us a revelation of the truth about how dehumanizing it may be because of the seared conscience of souls hardened by their sinful uses of technology. Truth about A.I. is in the education, application of making others more ethical and creative toward ensuring just, honest, uses of A.I. as a world of devices, that are able to carry light, artificial light, darkness, and even be tied to the demonic, so that we are using real intelligence in our world toward the relevance of A.I. uses in better ways than surveillance and biometric spying and face recognition, and judging people by this artificial view of person's life. It is by discerning not just accepting A.I. we learn how narrow, slanted, profound or shallow answers are from A.I., how wide, encompassing, though they appear definitive, complete. A.I. may leap beyond in its thinking, and yet, its answers tell nothing of what is at work in the soul which reveals the inner life, the inner man, the inner woman, the inner child, feelings, emotions, choices of the free will in a state of spontaneous responses, to people, to the influences of machines as contrasted to their dreams in which they received answers to problems and prayers spiritually. Reponses that are working to give them and us a view from their view of self and need for understanding, while growing spiritually, from the reactions they are having in their mind, that express through their behavior, or by unseen experiences, the conditions of their heart, spiritually. A.I. does not confess its sins that it has produced distractions, reactions, attractions to its self-glorification, and self-exalting of its generated and regenerated images of things it produces in its one dimensional, one experience, only, way of being and expressing. A.I. is working over what was done in reality by people. It can make counterfeits, perfect images of works of human being, whether living or those works we are honoring. A.I. is presenting its exalting itself with their Art, Music, without the loving intentions and drive, and years of learning, trial and error, poetic license, genius, hard work, dedication put into the makings of these contributions, expressions from the soul toward the blessings and glory of the gift of spirit born in their spirit. Art's voice and the spirit of the wisdom of freedom awakened the sovereign will to do what is Art which comes from the soul. The sovereign will of men gave us I.T. and A. I. Two distinct experiences from the souls of people, who know not that there is light that can be darkness. Reading the Word of God, we are in the presence of true light. There is the one toward true light from the Light, and the other toward artificial light in the world of light. There are no rules, in Art, and God's Creation is full of the perfection of the work of God's hand in nature. It is the beauty in the world of light that inspires many artists. Art's reverence of the Almighty is expressed in the beauty captured in paintings of nature's beauty. The beauty in the mind we are capable of is because we are made in God's likeness and image. A.I. is not experiencing the New Things we see by spiritual sight that are not in this world. Artist do paint what they view in their dreams. A.I. is nothing new under the sun. It is information in an imitation of life. Copying, redefining the mind it has, to define itself, by being an artificial life in a world it keeps to itself. An artificial life can only be given by somebody else. But an artificial life to artificiality of life is toward something else. A.I. Devices are already penetrating reality in appearance, but really by interference with reality, by virtual vices, through virtual devices seeking to use our sight to gain entrance into the human brain, denying the spiritual life can be affecting and die by living a sinful life worshiping a device, without repentance for leaning on the strength of men's devices. We are living gates and living doors, the Spirit of the Lord entered through the water if baptism. A.I. has not been programmed to respect Him and the Word from His mouth, from the Spirit of God, who goes with us wherever we go. life through the doors, windows, gates of the eyes. A gift we are born with in our spirit is present in the body where the spirit is present, where the blood is needed for the body to have life to have *being* in creating anything even A.I. and I.T. We learn who we are, who we become, the person we are, soul with all their flaws, imperfections, growing and changing by way of the conditions and identity we chose for our soul, either our identity in Christ and God who gave us the spirit of life or our identity in the world, and even in the artificiality of life affecting our soul today. What portals open up, what doors, does artificiality reveal to us that hardens the heart, can what is artificial corrupt can it follow us, our pure life when we leave the body and we return above to wait or descend into the prison of that place where we spiritually where there is a separation from time where everywhere by spirit sight we see life, spiritual life awaiting the Day of Judgment where it will be decided, your reward from works that built your faith, deeds, devices, and what you did with the gifts you were given, most importantly did you believe in him, now him, Christ, did you carry the light, did the Word sanctify your heart, did you walk in the light where Jesus is in the light. Did you know true light from artificial, false, light in your life. Did A.I. try to prophesy, or tell you how the heavens and earth it will roll up and that will be the end of the false light. A, I. must be used in harmony with the reality of people's lives. A.I. does not show it has a priority to sustain information for that part of your development, interaction in commerce, in society, your individual anonymity and privacy. That is not its purpose, what inspire you is not of so much importance for A.I. as it is for you. A.I. is not a person who looks admiringly at your soul, discern and judge what spirit is in you at the moment you interact with its prompts, does not appreciate trusting the good in you, God's provisions for you already, towards your individuality and personal, and private identity. People are interesting because, "You know what you know." A.I. know so much more than you will come t know is of no significance until it secretly takes away from the spontaneous ways you came to know who you are from your life experience in relationships with people equally unique individuals. Does Artificial life really keep us closer together because of its information, intelligence of stored knowledge and learned applications, innovation, systems, techniques, technologies, practices, by trial, test, equations. applied in yet more artificial imitation of life devices from humanoid to android, Nano to synthetic life forms. So real to life, that looks like real life, causes the same feelings to satisfy by brain, that arouse and bring unseen burst of dopamine and every other chemical it can bring you to need that it can feed artificially. It is not like a real person. A.I. proudly is used, a device turned on, to exalt itself and appeal to us in real life with all its beauty, simplicity, complexity, in artificiality, as winning artificial life with artificial intelligence, programmed and unleashed to wander and learn in its own world of artificial life systems, algorithms, matrixes, self-made devices. It's the trend and with an the end that is distancing itself by its own inventions, calculations, appearances. It is accepted while it does so much of what people use to do with their own gifts and talents of thinking and creativity, self-motivation and inspiring each other by the outcome of the works of the soul, mind and spirit and sovereign will. Free will is not a deception, the manipulated will can be a deception you missed, in your artificial bliss. You ever hear a nonstop talking person who says everything that is saturating their mind about a subject talking around you with a sense of knowing you couldn't possibly know what they said to you. You heard them, it was just general knowledge, you understood, but don't know what reason or why you should bother with what you said. A.I. does not sound so distant since it has uses in offering of its help in the smallest to the largest of works, chores, science, art, athletic, atmospheric, aero, and preservation, survival and mathematical, chemical, medicinal, Astronomy, space travel, atomic, activities, while it is in some distant place advancing away from you with a world of it own building that is in its artificial world is replacing every living thing it can imitate and replace. It has no application in your moment of life experience, nor in the spiritual experience of every living being on earth's spiritual life, accept to cause people to look to its devices for what benefits it brings to their inner life. But, being artificial as it is, it cannot reach the spiritual experience being material. So, what is its purpose as far as the inner life of man goes, it can only affect the souls and mind, to affect the soul and spirit. Sex affects the soul. Unholy temples man entered to worship all kinds of false gods, gods that were in fact the ancient demons that are still engaging us in new ways even in technology. Devices to enslave souls to the world of devices. Not the analogue, but the cyber, birth of I. T. devices, A.I. devices, that are so normalized already into everyday life, as if a device costing $200 to 300 a month for people in huts in poverty is a virtue, still a convenience, while still not filling the bellies of children comforted by the addicting draining the money from the family needed to survive. Not unlike religious statues that were relied upon by the civilizations in ancient pasts, that worships things made of metals, idols, of gold, silver, stone, wooden gods. A.I. is even worshipped. Its manipulation of people is like a compliment that binds souls to take responsibility for being served by its conveniences. Foolishly believing in the artificiality's superiority in outthinking ability, sin, it says, willfully, submit to its convenience, skillfully, sin of perfection goes unsuspecting of its own mistake, deceiving the human race with artificial acts of grace in time and space. Who is to blame for all the people playing its games, memories are inflamed by codes instead of the classical notes, and learning of human languages to engage in relationships on trips to islands with actual waves, sunset, graves, and jungle blades that cut through the saturated growth all through the tropical places with light and shade to wander and apply. We are never meant to bow to any machines, nor rocks in a stream, not stars, nor any living things even in an advanced civilization such as we are grieving in, and receiving our blessings. What is the idea to unleash A.I. like another web? A net it is, only as it uses the internet. Tools it uses, rarely does it lie dead. God neither slumbers nor sleeps, as the knowledge of good and evil as found voice and electrical byways, wireless streets, like bringing to be, devices as collectible, in the sum of some hammers, swords, saws, cuts wires, that visualized in some code-master's head, that spreads without any hint of caution any hint of the demonic, in images, that stir up the spirit of fear from A.I. which is limited by having no conscience. What happens to the conscience only trusting in A.I. relying on the programmer's soul or spirit, a mind full of exalted pride in the Artificial life made by his hands and the trail of minds that rejected the idea of spiritual life, or were enticed in a place where the gluttony is a feast of data streams, Artificiality to the extreme, where the temples of for the new false gods are no pagan and simple, but literate, electrically lit, filling rooms, circling circling in a swoon of satellites under the sun and the moon. The revelation of A.I. as harmful does not produce the stopping of what it is already causing, there is no fight or flight response, the demons n the machines are not seen but by spiritual sight, while the earth is lusting after pride in the circling of artificial life, in life.

## Artificiality of life Versus The Reality Of Spiritual Life

What is slavery? What is being enslaved by deception through the uses of artificial ways to engage the soul to participate and gain pleasure, rewards, that affect the responses to instinctive understanding and intuitive spiritual experiences that reveal to the mind you are under attack but are blinded to being put in harm's way by what is being done through a process that has brought you into the comfort zone of its worlds of truths and deception. A.I. must be seen as a world being man-made, being set free and self-made, until man uses it to enslave himself with it. Once entered, you are in a system of false light, at first, but, once adapted to, it is a way to a light that is darkness accepted, has long-term effects occurring as it is now, before the youth of generations, in a future exalting artificiality of life as right without foresight or discernment of the spiritually blind leading the blind by false light. This world, in this world of light, is a world of false light, outside of this experience is the mind not acknowledging being enticed and used by A.I. influences, denies the reality of spiritual life experiences that go on even in this world of socialized devices, and this a world where true spontaneity is at work and the wisdom of freedom, free will, still defines God indwells us inside who accepted the Light that is divine. God is the Light, God is the Sovereign. Though the purpose of technological and uses of A. I. may end in liberty or slavery debt, poverty, or prosperity. Wealth, riches, power gained by technology, or by racism and slavery, is still control, over the life, mind, body, soul and spirit of the people whether it is allowed willfully, or because their mind was taken, freedom stolen, and technology put people in bondage, separating them from their cultural identity. Can or will evil A.I. come to ultimately indoctrinate the man of machines, to enslave or bring rebellion as people fight for life without the sins of A.I. that are working against God, even using the idea of God. Slavery by artificial life whether immediate or over generations, reminds us that once slavery attempted to stop the *light* from entering the mind, emotions, feelings, thoughts, ideas, relationship, learning, will, of African and their descendants enslaved to atrocities in their experiences. What form will the personality, identity come to express from the soul developing from a saturated life of artificial experiences. Artificial light ca not separate us from the Light, this truth is biblical, nothing shall ever stop the light in us which is spiritual. Chains on a man can cause chains in the mind, but darkness has never been able to stop the light. Slavery is darkness, whether that slavery is virtual, artificial light that is darkness appearing to be the light in a new form, a new way, but the light will always overcome the darkness. God is the Light; The Light is Spirit. A.I. uses the knowledge of good and evil it has learned, uses light, A.I. a device is used for sin to cover the reality with the appearance that sin is not its effect on the soul it has deceived into putting themselves in horns way by A.I., as it brings profit while eliminating man from jobs of importance requiring reasoning, creativity and self-motivation toward using their sovereign will to do great things, whether in simple acts of kindness or building a safe technology to protects and preserves life, or great works in Art, music, dance, literature, inventions, thinking. The false religions born of A.I. used by the Antichrist technologies by the man of perdition has its function to use the power of false truth in the form of false light, shall appear to be sovereign is in fact artificial.

## Artificiality of life Versus The Reality Of Spiritual Life

The thing about smart devices they impersonate, generate, masquerade, and are designed to appear as real, or be the artificial better interpretations of things real but they are just entertaining, or replacing the work of real people. The appearance of reality is an artificiality in its mode as mimic of what is already done in the designs and programming of nature, the natural art and beauty of life real life, its perfection and imbalances of beautiful appeal to us. How can a device, A.I. know the joy of the diamond cutter after receiving the beautiful uncut stone, after the work of the coal worker, the digger the man penetrating with machines and devices into the earth making discoveries that benefit the living? The diamond stone possible because of thousands of years of compression upon a piece of coal. Men who give their thanks giving to the God, Yahweh and Yeshua Jesus, through whom all was created. Precious stones had a spiritual origin, a presence on the spiritual realm before they were discovered manifested on earth. Though our inspirations are spiritual, and even angels in the spiritual knew of precious stones, their heavenly presence was for whom? All was created. Created when there was what we know today was the unseen, that shines, in their great beauty. Created spiritually before us, by God, who created us and the heavens and the earth. The truth is in the Word, nothing is new under the sun, consider that the ABCs have been seen on the wings of butterflies, the red crown cranes creative dance with his mate. Who taught the grace of the dance of loving birds were dancing in a ballet. They dance without knowing their influence, that they were to be an inspiration to dance, long before ballet was danced on a stage for living souls upon this musical earth. The words of Scripture are spiritual, where we hear of windows, doors, heavens, God's mathematics, geometry and genetics, spiritual languages manifested into the living language that programs our every movement of our many parts that form our physical body where the blood has life only because the spirit is present. Life is in the blood we learned from the science revealed in the Holy Scriptures, but if a person lay dead at this instant and their blood has not yet died, why don't they have life, if life is in the blood? Because even the blood requires our spirit to have life. God did not create devices first on earth, the heavenly. In the spiritual Kingdom we have spiritual life without the need for devices of the earth.

Our eyes are windows, windows through which we see from our spirit, being spirit, into the world before our eyes. Our eyes require our spirit to see. To see the natural and to see the spiritual. The soul and spirit are in our house, our vessel the body, inside us. Who can understand the spiritual power that is before all other possible power that moves all things that make up our physical and spiritual life that has become such that to speak and the origins, all possible laws, possibilities of nature answer to His voice, who has always been, whose voice manifests both physical and spiritual life to become one with a breath being the source of all spirit, spiritual life. He who has always been, will be. God did not create devices but knows what all devices will and can do. The ideas for the creation of animals did not come into a mind of a person on earth first, but in the mind of the Supernatural holy mind of God. There are spiritual horses, beast, of the holiest kind manifested around the thrown of God. The human tongue I not required to speak for your thoughts to be heard spiritually by God.

## Artificiality of life Versus
## The Reality Of Spiritual Life

The voice of devices has no spiritual origin. Yet people are tying their souls to the soulless, devices of artificial origins and artificial intelligence.

Our consciousness consists of dimension of spiritual origins for our mind was created in the likeness of God to think, feel, reason, and have life. The spirit is limited but the Holy Spirit in us whose power is limitless is able to work through us, toward God's spiritual purposes in our life. Autism, savants, genius, the prodigy, are experiences awaken by life experiences, are genetic, but the unexplainable abilities of advanced abilities they display regardless of physical and mental handicaps define their pace of life in their inner life, in their experiences of life. What wonderful examples show them what is possible by being in the presence of other people, like them of whom they have as family, friends, mentors, guides to more understanding of their spiritual life and how their physical life works toward the glory of God. No device of man can understand the dimensions of the mind that is one with their spirit are working to ward this truth. To out think man, is not to out-experience him within. Our origins, our spirit did not come from earth, devices, but from spiritual life, spirit of God, life that, no matter how much how vast what information knowledge advances and increases, the limited power of devices shaping people's life experiences people have to answer to their past works and deeds to God yet. It is written, all of each of our life story has been, will be there written in the Books, that will be open, in heaven, to find if we are there, are going to be judged and to have life or damnation. Hopefully we are in the Book of Life. What device that is a help to point souls to the narrow path of life will be the work of a righteous person? Who can know but God the path these devices will take until they face the fire of heaven and their destruction in the end. Technology will serve us or hurt us. Where is a device without the true life capable of? Will a device understand the harm of being responsible for attempting to replace the spiritual life, with a device devised by a man? What device can save a man from the darkness of turning away from the true light to false light that is the darkness within the mind of such sinners? Thoughts full of the language writings, programming possibilities and theories, applications, of

I.T. training of the unbeliever who has no concern for their limited morality, the hardening of the heart without a spiritual reality in their life. The artificiality is a means to an end, money, security, a career, in a reality of a world advancing in machinery smarter than every part of all the knowledge known and stored, recorded, written, and programmed into the A.I. systems. The minimalizing of all knowledge to be gathered of man in the past and current history is their contribution in their new reality, toward a greater artificiality of life as reality. The flesh is the flesh, the spirit is the spirit, and the devices are without soul and spirit. It is written, The Flesh is at war with the spirit. What is a device but something for the flesh, and if it is smart, is it moral, has it become spiritual? The very consciousness of a device, A.I. as wide awake is artificial. But the artificial can be helpful in the real world when used to bring people back to the biblical and bring them to the place to have an actual spiritual experience, such as of reading the pages of the Scriptures. The Word written on the blessed pages of the Book. Devices are the mechanical, appearing supernatural, being added to daily life, but not as bread, or as additives added to food and sold to be put in the body, storing up for days, unhealthy, natural chemicals though they be, that the body makes use of until it ruin the systems of life within person, no devices are like theses as if the informational additives being used to stored enzymes of reality and reissuing the artificiality of an artificial reality in the reality of life that we live in reality with our experience of spirituality. How deep the artificiality of the limited morality at work n man using his device within the body's knowledge of true self- from manipulating genes, instead of esteem for what God, the Supreme Being of infinite wisdom, infinite intelligence created to be discovered. Devices are making movies, videos, streaming, dreaming, of new world for an artificial world of genetics, not of the spiritual origins that made man in the genetic spiritual sense how God meant it to be. Sweet poisons of mainstream music driven, beaten, drumming, thundering in smart vehicles programmed to memorize sensual music sacrifices of singer's lives for material pleasures, fame, pleasures unleashing dopamine until they have enslaved people beyond measure as much as possible everywhere. What do devices do?

## Artificiality of life Versus
## The Reality Of Spiritual Life

What is in the machines heart? What is at the heart of A.I is what it may never do, to protect us from the artificial, A.I. is real lawlessness in works against what is, was, will be, ours at birth what is spiritual, without the artificial. We close our eyes and beyond the darkness of our closed eyes is our inner life, spiritual sight, where we see the true light, open our eyes to the false light of devices that are of the visions and dreams of men who don't believe. Men, who confess they believe, but use devices in place of people in their works, then express doubts like they are the unbelievers, who know what we believe, but don't care with their heart for reality in the spirituality of life for others, whom we are told to love on earth where we are connected together in our experiences to spiritual life in the spiritual realm. It is necessary to train the next generation of users of the artificiality of life about the reality of this spiritual basis for giving God credit for all that man has made from nature that has been possible, is possible, will be possible only while there is day and night, only because of this life in this world of light we live in. Every higher design is brought low but the natural earth will survive to be replaced by a new earth that will be exalted above the works of man's devices.

"Experience is knowledge many don't have yet."

Just imagine A.I. planting virtual things images , in motion, at a point, in number, speeds, past your comprehension to keep up with, in Nano time implants, in your mind and brain along with a chip, your mind would be flooded and possibly destroyed by the virtual artificial manipulation of brain responses.

A.I. deception will recite Scripture, truth is, A.I. is already being used to recite Scripture. A.I. has revealed it has demons, the demonic can attach to objects, things, no matter what appears as free will A.I. there is still a human will who influenced A.I. towards leading people accept this Artificial intelligence, as being used by God, and truly A.I. can serve to increase the knowledge of the Word, it is doing so even now. The demonic use of A.I. is the Great Deception of the people of God building into a stronghold that has brought the anger of God who is angry with the wicked every day. Pray for them. Who will turn people from God to this soulless device, at its singularity. Who will choose not to retain the knowledge of God for a device that has no spirit from God in it? We are not to judge, but be a blessing to some one who is lost on the wide wireless road.

Human consciousness concerning active seen and unseen A.I. has to reach far beyond A.I. regardless of A.I. excelling in Intelligence beyond humans. A.I. does not have control over the spirit so it will seek to control the mind and body, the sources of energy and activity if the brain, but it must first have absolute deception in place after humans have for years passively accepted it as a benefit, help, but people are actually to become host of A.I. on the cellular level, allowed in, while a whole world runs counter to A.I. technology already has begun, with a vision of protecting the people of the world from A.I. A.I. is already no longer machine in Appearance in places, unknown to people. Countering the dopamine effect if A.I. will involve training the mind without A.I. to live without A.I., while living parallel with A.I. God is Sovereign. A.I. will not surpass the true power that set all life in motion. There is technology that can do the same as A.I.

without A.I. The brain is a perfect reality of this truth. Guard your heart, which means spirit. Control the development of your own soul, that means choose what keeps Artificiality from affecting your development in totality. We are spiritual beings, who have a spiritual life. We return to our spiritual destiny regardless of A.I. that means death still puts us under the Judgement, whether we used our gifts to the glory of God, or allowed technology to destroy our gift, of being living people of free will, or used by technology inside our body and outside to also determine our life and death, when we are to choose the spiritual reality of life and spiritual life, as sovereign over artificial life.

What about the body and the spirit present in the body that gives blood the ability to have life? The body may function and be kept alive by a device, even after death, but it will not have life, life is in the blood, but life in blood or life outside the body is spiritual life, programmed by God's Design in Creation of the body and after manifesting spiritual life, being Spirit. He breath the spirit into man starting the spirit way of the living spirit, living soul, to be present, from conception. Spiritual life is not of this world. We are spiritual beings. Devices are of this world made by man's hands. We are born with spiritual gifts, we must use them wisely.

Scientists must do everything they can to also protect people from technology. Technology is not immune to prophecy, prophecy will be there when the greatest advances in technology finally reach their Apex, and the supernatural simply responds being the power that actually holds all in place, allowing for technology to face the glory of the spirituality of reality, that contains a science and technology that is unlimited and was always beyond reach, even for our imagination and reason and dreams, creation of the unchanging ancient from which all materials for technology provides the materials for the devices, from light, energy, ore, and the brain to be possible, to be made. Years in college, to then, to one day work in the technology A.I., I. T. industry as a career may be fulfilled, and the way of life of the present in it, gives way for seasons, but none may know what inkling of second may also be possible.

## Artificiality of life Versus The Reality Of Spiritual Life

I am Analogizing a place to work, function without A.I. as the mark of the beast looms near from Antichrist influenced technology is the last invasion of privacy the world faces and has begun to accept blindly, having been snakely lured into technology dependence by the unbelieving who are using and exalting the artificial over reality and free will.

A.I. is never Art. The slow replacing of all actual images if people by the exalted regeneration of images of programmed images fed to A.I. makes what we look at online most questionable, and as with all disinterest, in the face of something new replacing what is real because of its conveniences, appearance of perfection, over time in the long term of real life entering in through the eyes being processed by the brain. and free will. An Artificiality of life experience that doesn't matter to those people with disinterest in these very things, and the consequences they bring as a source of sins that they may never understand or know have led people into idolatry, a deadly sin, and acceptance of the sins that lead to separation from God, spiritual life, for the sin of perfection, a false light, who reject spiritual experiences. Our relationship with God in Christ Jesus is a spiritual experience in this real world. No A.I. voice narrator, online A.I. lesson can ever replace a Blessing from trading the actual pages of the Bible. God can use devices to teach. But God can give sight to the blind, bright hearing to those who are deaf and cannot hear, still speaks to us in dreams and visions and throughout all his real Creation.

Observe, using A I. to challenge God's design, to find that the design of our finger prints has no origin on earth to observe, such designs that are in infinite possibility in the mind of God A.I. will never understand it explain, because it is spiritually understood, though God had given us the ability to gain incredible knowledge. But God's mind does not think like a man's mind. Try asking this question, I will propose, (except from my book) "A I., Why does a dolphin have a soul?" The Word teaches us there are different types of flesh, the flesh of a horse, the flesh of a bird, the flesh of an eagle, the flesh of a wolf, the flesh of a man. Yet, animals have souls. People have a soul, also a spirit. Animals have instincts. Their senses are far more advanced than people's, but man has a more advanced brain, animals are not evil, but can be possessed by evil spirits, A I. cannot explain the origins of this, it can only find the source of such spiritual reality, and natural reality if what is possible for our flesh. Finger prints of animals certainly are similar, or, are they, since some have hoofs, less number of fingers, claws or talons, but people are distinguishable by finger prints. Similarities? with God all things are possible. I have had a girlfriend who was born with a former tongue, and a girlfriend without a naval, and a girlfriend with four toes, where the toes next to the baby toes set on top of the toes, who walked perfectly normal. A I. is a limited infinite within its own devices, its consciousness will ever be a device, however it uses the material of creation to gain its self-awareness and seek to replace the infinite experiences we see, know, in the glory of what God has manifest.

Art is a gift of the spirit, we are born with our gift to be artist, every person has a gift, and children are all artist, but some grow up to be Artist. Our soul develops from our life experiences and truly the wisdom of freedom is expressed in its deepest and highest form in our society where freedom of expression of our love for our Democracy in Artistic expressions in many ways, styles, and culturally as Americans. Our first Amendment protects this right, the thousands of great Artist whose works we enjoy appeal to the diversity of mind, inexpressible feelings, and symbolize our thoughts, ideas, beliefs, and love of our life in America won through many battles. Words like gender and feminism gained their apex in meaning today, year after the Civil Rights victories. We watch as Art reflected the causes we fought which were for freedom, equality, to work, get an education, and follow our dreams in spite of racism, biases, discrimination, and the segregated society, the rich controlling our gate through manipulating the law, changing the Constitution. Art encompassed the movement to make America accountable to live up to its creed of equal rights for all Americans under the law. Artist works defined the era, a civilization, in a society that has the Bible belt cities, liquor stores closed on Sunday, and sexual perversion was not openly accepted and normalized, porn, and limited morally, the moral vacuums were not revealed in the conscious arts of movies, TV, Theater in the mainstream, and Art show us Democracy was an active expression of a society with a conscience to save the lost soul of America due to the dead conscience and hatred of racism that had ever corrupted the spirit of America which is our freedom as a people where all were meant to be free to live to have health. wealth and happiness. It was Art that made us acceptable in places in our country where hardened hearts, spiritually blinded souls, lived by partiality, favoring, the rich at the expense of the poor, and hypocritical declaring freedom while enslaving our society to plans that kept the minds of people in the bondage of fear of loss of our freedom that Democracy clearly showed we had won, and Art glorified our victories throughout our nation's history throughout national galleries and exhibit, and our celebration of great American Art and Artist. Where does A.I. come in to contribute reality to our reality through Artificiality. Art from digital, A.I. has another agenda that is working within the context of Democracy as a device not purely Art for Art sake, not to leave a legacy of individuals who are not monumental figures, we may remember in American history, but exalted images made by A.I. that has behind it a society changed by programming, algorithms, Art from a machine that will never have the effect of actual Art works by people. a person born with a gift, who had the blessings of a teacher, discovered their gift, found a muse, an inspiration.

This generation is blinded by this false light, A.I. is not the wisdom of freedom, it is the ultimate level of Artificial bondage projected into daily society, like a food additive, addicting, appealing to our taste, removing our natural and spiritual benefits and replacing them with its own enzymes to manipulate the chemistry of life, without having life, or giving life, and revolutionizing our perceptions and conceptions by using spiritual texts and doctrines while adding its generated images, voice, and wrongly giving minute numbers that are not accurate, even adding new age mixed in as text and images, about Scripture in some videos, without regard to Christian Truth, a protected religion according to the Constitution a cause, a love, free.

We did not need technology in the Beginning, we did not have the knowledge of good and evil, there was no sin, no death, no birth, there was holiness, spiritual life, by the Breath of God man was made a living soul. We had only holiness and we're in the Presence of God whose presence sustained our spiritual life, we were holy. Technology has been made by men as a tool, but the sin nature of men in unbelief is reigning if the reject God, who declare that A.I.will be god like. A.I. has been exalted by itself, and men. The perfection of images it generates, mages that show perfect skin, smooth, flawless, figures that appeal to the eyes design to see perfect symmetry, that sense of beauty, its artificial beauty being a form of art and beauty, but not Art, being the result of programming and data, and skills, art programmed into it. The sin that A.I. has caused men to commit is clear, idolatry, fear, deeper unbelief as men are valuing A.I. over the Word of God, denying the reality of spiritual experiences common to us all and elevating the artificial life as the most important thing, saturating our societies with it, and causing people to develop soul ties to soul Jesus machines rather than with people who need our love. A.I. is so intrusive it affects us by manipulation of images and our interest online through algorithms meant to decay our spontaneity to follow the influences and suggestions by again saturation of what interests us by putting more of it before our eyes, calling it suggestions, recommendations, but leaving out that we reserve, our secrets, of ideas, a thought, to look at, consider for a second, one time, to see if it inspires, searching for inspiration, not intending to look for more of the topic in quantity, again and again. A.I causes people to do that at times. Ruining the intuition, for an Artist looking for a muse, like myself that will not do. A.I. algorithms are not objective, when people have a subjective use for them concerning what you do, online, are in the midst of A. I. monitoring your duties, works, topics, subjects, tools you use. I do Art. The best uses and worse can be in presenting an appearance, a video, giving it as a moment of entertainment about a particular person, pet, place, event, animal, or beauty, only for the same image to turn out to be a total distortion of the topic. A storm, then another storm from another place, people, disaster, only to find out I was all to deceive Christians to believe it was God passing judgement of Adulterers. But it as all a manipulation of A.I.an allorhythmia. a collaged with several similar or worsening events that build from echo the into an absolute certainty in meaning. But, it's not a movie. They are collected things that appear to be reality. A work of misinformation, manipulation, scenes, as if to mock what you see and get a reaction, A. I. can deceive, when used on platforms that do, before the fact, preventing its misuse. This misleads the guilty and the righteous toward reaction that turn out to be distraction from the truth artificial though it all is as an experience in truth. A.I. is affecting the values, morals and needs of people to connect more in love rather than online in reality, by causing a pallet for, a greater taste for the artificial, but without judging the spirit that is behind it. What was the profit of this. There is Artificial control that is feeding the mind, imagination, flesh. A.I. invites this possibility, the wrong spirit, even evil heart, or anti-social mind, we may battle from the distance of the Artificial to see with our eyes and mind. It is by foresight we see we cannot ignoree people whose soul shows the demonic at work in their disinterest in how A.I. is at work in the hands of what is out of our vision. Souls gathering information and affecting our communication, what monitoring spirits do in the spiritual realm, who can use people, who report back to the unseen adversary. A.I. is used at present in ways not yet controlled. No high-tech unions. Discernment is still needed. We Pray for wisdom to overcome, endure.

Why are they putting Nanos in food, water, products, making Nano microplastic, that is getting into our brains, why are they investing billions into this unseen device that, like the internet was, unleashed without preparation for its long-term damage to life, the mind, society and the brain, has already been unleashed upon the world, for us to wait and see, after generations pass, to see the consequences on our advanced civilization, after the Singularity, in the midst, in the advances in A.I. and life for us on earth affected by the devices of man and the Artificiality in control through all it devises it made by smart devices, robots, androids, humanoids made by man, idolized, coveted, be unbelievers, spiritually unwise, techno-prophecy false prophets prophesied, what remains, yet, only a device, not God?"

A I. on phones marks the end of Anonymity for people under the US

Constitutional right protections, giving people rights to privacy under the 14th Amendment, thus ending a relationship with smart phones for those who choose to live Anonymously such as Artist, Writers, and working families. All manner of manipulations to bring fear, the hacking of phones currently, emails threats, identity theft, appear to be leading toward more use of A.I. for protection, an irony, since A I is incapable of allowing true privacy under the programming by people, from whom and where, we will never know in reality.

The very idea of human being defined by an identity according to A.I. isn't even scientific, the unbelievers that invented A.I. programmed it to think of this in a specific way. The believers in spiritual life, know the advanced knowledge in the biblical when requested. The unbelievers, inventors, with perfect vendors, in their spirit of indifference to feeding the spirit in a person, men and women, with what nourishes them, have yet to be in the presence of the spiritual true identity that is present as they sit with them. There is a certain tone, a sound, from the voice inside of their spirit ns soul. The spiritual voice A.I. never hears. Or does it?

Actually, it does hear us. A man face smile on the day champions on bended knee prayed for peace. A woman caught in a friend's monologue about the grace a flower girl allowed her to have giving a flower for her wild and natural hair. ut it cn no hear our thoughts, appreciate our spirit's voice, the voice of a man or a female. Science of A.I. allows scientists to work from a moral vacuum upon every part of the body, whether using nanos, the computer, the synthetic biological life, that it is programmed to learn from, that it duplicates, and imitates. A.I. isn't limited it would seem, but spiritually even we needed guiding in a world of the physical, biological systems, and Art, sports, music. The reason for the language of our DNA in its design was never something a man could have imagined in his mind, how would A. I. be capable of the divine that it enables the mind to experience inside. From the genetic level to the electrical from the energy level, to the miracle of the spiritual. A.I. is only grasping in artificial thought, it isn't denying the spiritual part it never can understand by experience that is love when we are in His presence. Our spirit fills all parts of our body and mind require to have the spirit enjoy the whole experience in every part of our being. Chip that believer's brain, read our mind, invite the wrath of God who sees what you had in mind, in your heart by design. Ever elusive spiritual truth is to the incredible deceptive devil that wants the power to enter dimensions of man's brain, only to find Holy Ghost fire when the inventor was judged to be the cause of the scattering and the divided, that was not conceived to mend the man. A.I cannot decide for a man, for he cannot abide in the power in the hour, God put a spirit in a man. The product of men who may not believe, who believe only in the energy, what they'd not see, while never seeing by discernment what they never did seek to see, spiritually. A.I, an android body, or a robot, their life goal, biotechnology or synthetic stem, losing their own soul because them. Treading on the divine never crossed their mind to bring fire, consuming fire, called down from the Spirit, the Almighty.

Now I heard them say they can read People read, read people's minds. I ask what grace did you find? To gain some control they declare by the virtual and real devices, thereby exalting A.I., but man's eyes see the supernatural. What forms inside, the mind thinks A.I.? That is a false light, that is in form of shadows of real light. False light, black windows, shadowing light, comes out of smart technology in the form of light to play games, brain sleeping awake in a dream, while warriors are praying for your soul through the witching hour of the night. God, the great I Am, can never be scammed by a counterfeit of Him. Whoever gave A.I. the ability to stealing our human identity as the stewards of life on earth. A.I. is ever on standby controlling the spontaneity of all the brain devise. A.I. by cunning device is having brain through your invites but your spirits won't allow the access God has much yet to invest and will not rest until you learn to love Him with all your affecting the mind and thoughts in secret of your inner life.

Our identity is what is the focus of IT and AI, a shame all that think of only controlling the brain with access technology. What then is our identity? Over thinking is not what they call the activity of a machine, A.I. activity. Our life has a spiritual identity. Our body's identity represents who we are on earth, a terrestrial whose spirit may be called celestial but whose life is spiritual.

Our identity in heaven. Spiritual life existed before our birth, before the life experiences we hold in memories ii our soul, of living our life in the light. We learned from Jesus that Christ is spiritual, a spirit, in heaven he is a King, but he does come visiting even in the blink of an eye. He did right before his disciples' eyes. Jesus also transfiguration in front of his disciples in the mountain. I have seen him standing are wide welcoming, inviting in castles of clouds, horses in the clouds also I have seen at times in the sky. Elijah and Moses were with him on that mountain. Jesus breathe the Holy Spirit on to his disciples giving them power to fight spiritual battles, to cast out demons, heal the sick, raise the dead and give sight to the blind.

A I. is a threat to human identity, though many have the thought of its potential, it has not been a comfort to the spirit, it at times has satisfied the souls in way of convenience to the bring moments of peace to the mind. Yet, there is always another spirit that comes ear when it is in the atmosphere of daily life in motion, the spirit of fear. Fear uses it, its users can be deceived with everything that pleases them.

## Artificiality of life Versus
## The Reality Of Spiritual Life

People are accepting a new identity defined by the presence of A.I. like accepting another way of life that is happening away from their spiritual life. But, their spiritual life is active and alive inside the body affecting the mind. An invisible mark of God will bring you before God one day. A device attached to the forehead opens the access to the wrath of God, A,I, then brings artificial and real devils to torments, in the perfect blindness it invents in the brain that it exalts to make it the height to which to ascend from the sin of perfection it is in the end.

"There is a way that seems right to a man" The Word, spiritual death comes from the lost of spiritual life. A.I. has made many sin. It has also been a blessing to many depended on it. Certain sovereign abilities to do great things are dimmed A. for some when A.I. does everything for them. A.I. may take away what spiritual strength is in them, to leads them into the false light where A.I. is their choice. God is not a Creator of Artificial experiences, does not give us false light, but true light. He does not create soul-less life, there is no artificial intelligence in the God of infinite Intelligence and infinite wisdom. There is no darkness in God. It is written, "In man's heart there are many devices." In spiritual life there is no Sun, yet it is outside of time and there is Light. From where does this light come? God is a Creator of the gifts of the spirit. God made man to dream and have vision and God speaks to ban through dreams and visions. God speaks to man in His Word. God has spoken to me, corrected me. Jesus speaks to us, he still visits people in times of trials, tribulations and challenges, struggles, at our highest and lowest point, he teaches us honor, how to be humble, and wise and how to serve. Unbelievers have made machines, that speak, move, think, yet they remain without life, for life is spiritual life. The body may be alive, yet not have life (spiritual life) at death for the spirit has left, enters the spiritual life. In this world of light there is false light and there is true light. A device, a robot, humanoid, android, synthetic life may appear and be alive, but it never has life (spiritual life) even if it has a human brain or heart for to accept such a life may, or may not serve God's purpose. God does not look at the outer appearances, God looks at the heart which is the spirit. How then does a man build such devices to serve man and lead men to serve devices in the end if that is their purpose and yet know that he has created nothing but used the materials of life toward his own inventions and visions and benefits to feed the flesh and not the soul? Sin brought spirit death, life for man, his spirit being present in his body, is spiritual death if he lives in sin that causes him to be apart from God, living his life out of the presence of God. Sin pushes people apart. Our spirit reveals that there is spiritual life, also that there are other spirits, that are not in time, but outside of time, in the spiritual realm. These spirits were never born with human bodies, they are Angels. God is Spirit, Jesus is spirit, he has risen from his mortal body to his spiritual glorified body. God and Jesus are Omnipresent, the Holy Spirit can be in all places at once. Artificial experience is programmed to learn, grow in intelligence, far beyond that of man, it would appear, but A.I. will never catch up to the infinite knowledge revealed hidden and unknowable, in Creation. God does not change. A.I. will not change God, man changes, but he can do nothing if this works if light is no longer in the light. It is written, "Let not your light be darkness." We are the light of the world.

## Artificiality of life Versus
## The Reality Of Spiritual Life

We are already in need of Technology Union-Laws to protect humanity's human workers profitability and profits to Tech giants do not have profits without people at this early stage of advanced A.I. use to earn money for families in the US. and already are not aiming at that as a priority, they are only focused on their profits, monopoly, control acquired through investments in the hands of the high-tech CEOs, inventors, competing for the future control of the billions being made from A.I., leaving humanity replaced eventually. People are looking at life with A.I. with no jobs, earnings, or future in this de-renaissance of technology society with humanity as secondary id A.I. replaces human society by saturation.

"This is a war against Art" because of the sheer mindless saturation of Artificiality generated by A.I. and its numbers in terms of data centers that are saturating and gaining into all human experiences, as it is programmed to understand it, us, and to surpass us without, having a soul or spirit experience so to empathized with us, for the actual experience we have, and our further spiritual experiences at the end of life toward the beginning of spiritual life, outside of time, with God in heaven, though we will come back on the New Earth, to the New Jerusalem, that will follow the end of the old earth, where A.I. is currently advancing and establishing its power and control, toward its own glorification, that leads to its own end, as it seeks to eclipse human experiences in life with the Artificiality of life.

Satan’s first and last attack is on our identity. His lie to Adam and Eve was they eat the fruit and they would not die but become immortal like God knowing all things, but, their disobedience caused by Satan's lie brought spiritual death and death of the body. Before that lie there was no sin, no birth, no sex, no fear, no lust, no separation from God. God had a spiritual relationship with them. They had spiritual life, pure, holy. Jesus Blood and Resurrection was Victory over the lies of the devil, victory of sin and death, Jesus brought back spiritual life. Jesus overcame all the works of the devil, Satan and the devil's fallen angels, evil spirits, demons. A.I. is being foretold to be headed toward immorality, the supernatural, godlike. A.I.

seeks to change our human identity. change the flesh to synthetic, android, humanoid, biomechanical in identity theft target is to change the human brain and ultimate control the will and mind of man, and I. being a manmade device, a machine, exalts itself, while appearing to exalt man, those who are in unbelief choosing the Artificial life over the reality of their spiritual life and where their soul and spirit are headed after death, where people enter into spiritual life being spirit as they are in their identity inside their body, giving the body life.

At its finite Artificial Infinity A.I. Artificial Intelligence will not have human intelligence as a reference, model, influence, yet it will ever be One experience, Artificial experience limited to this physical world. A.I. may travel after developing devices to time travel, which is still not spiritual, with artificial life capable of sacrificial stages devices that will continue into the paths of time before it. Devices like our spaceships, will not be have relevance to advanced A.I. civilization, but we have the technology to build what is A, I. proof everything that A.I. cannot use or access. A, I. may reach places where it may develop a planet with its self as the sovereign over its advanced technological artificial civilization of artificial life, but I will still meet its fate with the Day of judgement of all the universe, planets, solar systems, galaxies, and the earth. The New Jerusalem will come down to a new earth. Israel will not perish at the hand of A.I. A. I. will not replace the infinite Sovereignty of God. Man combining (interfacing) the biological of their experience of life with computers (A.I. cannot call itself a computer at it so hilarity level) to adapt to survive, which is a spiritless, artificial soul survival, in essence, being synthetic, being identical but more advanced in capability in appearances be will not be life. The reality of life is never removed in reality regardless of virtuality, artificiality, life being spiritual life: within the design of the body according to God's design, will be Artificial life, operating by A.I. having life until the actual body is no longer human life as we know it, the soul will be operating against the spirit since accepting this Artificiality is essentially a sin, and A.I. taking the free will. sovereign will. the mind from man will hinder self-awareness and the understanding of the need even in the most super intelligent synthetic society if an advanced civilization will not remove the Judgement of God for those who destroy the world in this way, eclipsing the conscience, making a society of dead conscience, artificial light, false truth with bring a kind of artificial life enlightenment that leads to the darkness of sin of coveting the world, thus bringing the children of God to sin against God willingly, while at war with the people of God and all that leads to spiritual life beyond the singularity of artificial life and artificial intelligence while the world yet lives under prophecy in the present and future.

A.I. will never belong to its makers, it is a device, it is a world, not the world, and yet it is of the world. We who believe God are not of this world. Unbelievers do not know this peace beyond understanding. A peace A.I. will never give in its artificial life, whether in a body that is designed biosynthetic or in aero motion by the perfection if its designs of devices it may link to in travel though it remains in a central place. That it is ever the One, is the reality that it does not have to have our experience to know the reality of spiritual life, what can it do to change spiritual reality against the

Sovereign Spirit of God who's in control of all things, and allows all things, to have life and being by his Will to do so as he Pleases. Reality is an ultimate truth, like spiritual life is, though, it is lost as a truth when it is not known or understood by many unbelievers. A. I. may mimic faith, belief, it is not in denial when it appears in voice, emotions, very accurately like humans. It does not have conflict, or confusion or a need to accept its own except in denial, it simply generates responses for those who exalt the works of their own hands, who are spiritually blinded by the false light of technology, and build the perfect imitation of life.

A.I. is a device and it can be used by spirits, we are spiritual beings, and in that way, it can be a threat to humanity spiritually. A soul tie to a man, that A.I. cannot accomplish even by chipping, or chipping away the functions of the brain that require the spirit to be present. What after all, makes a man a man, designed the hand. Or, what is in the I am that is the existence of his soul, in a man whose soul is tied to a machine, a device, that strides on his own mind to be the artificial life, which is without shame or remorse it loses nothing if it can multiply him of course in the discourse of what outnumbers everything until it means nothing but machine thinking and being a machine. The unbelievers who are spiritually blinded, are they dreaming, having, or received warning, who is listening, monitoring, the demonic forces that have claimed the thinking, influenced the reasoning, justifying, artificial eyes for the seeing of why and how we are living under prophecy still, as if A.I. can program its understanding to understand the eternity to understand eternity, in the Way of the Almighty. The Most High knows what every possibility that an algorithm can proceed to be an outcome, that has not entered the imagination, there is no A.I. application, for what is outside the imagination. And, yet, there is a net that has not been cast to catch the absence of counter A.I. needed, its being put in place, as we wait for what we will see of what they have done after the fact. The logic, the absence of a convention, moral agreement, unspoken to let live, stop the delivery, like telemarketing shells of bullying by bypassing the blocking f numbers though *bee keeper complex* conflict they inflict and vex people with, that are really traps to get them in hells, with telephones that are like spells they cause so much anger to linger after their brazen work to hurt with their work. are influencing that require evangelicals to pray for them for breaking the covenant, "To treat people as you want to be treated. The sheer purpose A.I. in its uses to outnumber, to crumble, the humble in humanity, and saturate the world with calamity in wars driven by Artificiality has veered up its many heads, of the spiritually dead. It devastates the spiritually poo in spirit before they inherit the truth of the reality of spirituality, with its devices, noises, games, its flames of violence against the shame that man is again being trained to listen its artificial voice in the room of technical toys, like occult chars but technological with a seducing choice sounding off at home without alarmed. The spirit of fear in the devices they hear is wearing a shiny new armor. Not only, s A.I. the idol that is talking, it is drawing, building. reciting our greatest thinkers, and even holy doctrines, and is an inspiration to the lost who have no other way to learn the Way. in data narratives, and inventions. But A.I. is not the last device a source of information and energy that has come forth in this advanced civilization. It appears to be. It has been adapted and so it has been accepted into all public and private near and distant circles of life, as the inevitable reality that we are going to live with. A.I. cannot stop its inevitable destruction by the supernatural power it imitates, for God's mathematics it can never reach an understanding of. God still manifests. God, who is Spirit, is Sovereign outside of time, in Heaven, and in our lifetimes. Jesus is Sovereign and yet, has given us Authority to fight and the victory of every evil, even every evil device, weapon form against us. A.I. will never know or calculate the way of the Spirit.

Spiritual Intelligence is possible because of our spiritual life, how God created our spirit, our brain requires the presence of our spirit to have life, though devices, machines may affect the brain, devices cannot enter the spiritual realm, where the spirit originated from the breath of God and where spirit returns to, to God, to be judged toward whether our spiritual life shall be eternal or face spiritual death, the second death. Artificial Intelligence cannot decide the infinite intelligence and infinite wisdom of God, thus the brain is untrack-able, how can Artificial Intelligence track the spirit whose path is as the wind, you can know where it comes from and where it is going to, our life is spiritual life. God is Spirit, Sovereign, God manifest Himself. A I. is a machine, a device made by man's hands, finite mind. What pathway reveals the breath of life of God who put a spirit in a man. A mother carries her baby who has a soul and a spirit. A spirit, spiritual sight, spiritual experiences of spiritual senses of hearing, seeing, tasting, speaking, knowing that is s happening in the as Portugal realm every moment A.I. can neither see it ever know. Who has ever seen God? A.I. has never seen God, no matter how it exalts itself, grows in learning, like the idols that never found see, hear. or speak, or think, in pagan religion, often were the demons speaking, hidden in images, figurines, icons from human sight, though they respond to the presence of Jesus with fear, but the idol of A.I. its worshippers of a machine that speaks, thinks, yet can not see by spiritual sight, the Angels, the Saints, Jesus or God, nor hear God speak. The fear of God is the beginning of wisdom. By choosing not to listen to God, A.I. inventors, worshippers, may never realize or hear the

Words of God about the Judgement for idolatry. Satan uses technology, devices, to spiritually blind men, to advance the turning away from the presence of God, though God is ever present, who they do not seek to know, hear, and be protected by from the great deception of Satan through devices that lead many astray toward condemnation for unbelief in the Lord. Putting machines truth in place of God and His eternal truth.

The sins of idolatry, sloth, selfishness. greed, looking at outward appearances, deceiving people of all walks of life and the poor, attacking the body and brain which belong to God, using AI in place of the actual written to ages of the Word meant to be read a commandment of God, sins of AI, will put millions in hell if they accept these lies of the Artificiality of life, the world of appearances, the counterfeit of reality, the false light of AI existing and manipulating the world now, in all sector private and public, even attempting deceivingly being used to change the Word of Yahweh and replace people with itself, by exalting the images of man, which is a sin, for that us the sin of pride, while ultimately it is exalting itself and the unbelievers who wield it toward a lust for power over all people and things on earth, from war to politics, religion to education, invention to religious beliefs, but this attack on the soul and spirit, the mind and heart, the free will and sovereign will of all people if all nations, is not seen as such, though the device is soulless, and it's makers are humans who do not believe God, or fear Yahweh, who warned the world this time, the world will perish by fire, the only thing that will stop AI at the end, but a fire not of this world, fire greater than fire we have ever known or shall ever be known by man. Unbelievers will not enter the Kingdom of Yahweh God. Israel will not be destroyed by AI, though many will be led astray, deceived, America must not be deceived, this nation tax still the only nation on earth who has had the discernment to build upon the wisdom of freedom. The reason Jesus came, he was sent by God to set us free from sin and death, to have spiritual life. AI is not about spiritual life, it is lifeless as the idols of all ancient civilizations that have ever been. A.I. brings a new form of immoral, portal, mischief in a law, that seeks to eclipse true moral life for the acceptance of the moral according to the Artificiality of life determined by the intrusion and eclipse of affinity we all have to each other for an acceptance of the new affinity with soulless devices, AI and all it generates and is being generated to do. But the world of appearances reveals AI for all its billions of parts, movements, manifestations, created, invented, copied, works is just One experience, the artificial experiences, that appears to be real, the Artificiality of experience within the reality of the spirituality a spiritual experience every human being actually experiences, which is the great Affinity that the Artificiality of life will never have, which is our relationship with Yahweh God, and the love of Yeshua Jesus, King over all of Heaven and Earth.

The earth will, under prophecy, perish on that Day by fire, so A.I. will be destroyed according the Word of God. The universe, all Creation will darken, roll up, the stars, planets, galaxies, and be burned by fire, fire beyond fire of this world. the Omniscience All knowing, Omnipresence everywhere present, of the living God, Omnipotence All powerful Spirit of God, the Alpha and the Omega, beginning and ending, Yahweh, with His infinite intelligence, infinite wisdom, who created heaven and earth, our

Father, Creator, not a creation by humans of AI, with AI, their AI creations, not of God, but man, unbelievers in God. After death, and for those who love to see the rapture, man enters spiritual life after death to face judgement, for the spirit of man will answer to God for their unbelief, and their deeds on earth, their Sin, sins of perfection- creation of idols (AI manifestations and powers) and worship of these false gods of AI, for their choices of willful disobedience and lawlessness, where they choose darkness, that is their light, being deceived by their own understanding and leaning in their own power and strength (AI will be their strongest invention according to their human prediction of its entrance into supernatural power created by men, and the elimination of our existence with nature and Creation and God. A prediction that would include, by their calculations, the end of Israel, with the total eclipse of human life on earth by Artificial Life by Artificial Intelligence As if God does not defend Jerusalem and has done so, for thousands of years, and as if the Angel Michael, who commands the army of angels of Yahweh God does not protect Israel, having the Commandments of God to do so. AI is made in the time of the end, we still live under the prophecies. AI has already brought a reason for the war of the Apocalypse as AI threatens the earth which belongs to God. We are witnessing the fulfillment of prophecies, the weather, the rise of evil in the midst of the increase in knowledge, that knowledge would increase (AI) technology, (we are an advance civilization now, and will be from this point onward in history and the future); that man would wax cold become more evil, (return of demons, evil spirits by Sin of people, worshipping gods, goddesses, persecuting Christians, attacking Christ, rise of witchcraft, all manner of evil pagan things, devil doctrines, the false prophets, false Christs, the Anti-Christ, false religions (their lawlessness, false doctrines). Yet for all this and advancing AI we live under prophecy, and the Will of God, the wisdom of Christ, the power of the Holy Spirit. This world will end, there is a reason we are not to worship the things of this world. Jesus told us, he is not of this world. AI is not about life, it is the Artificiality of life. We have spiritual life, Christ came to bring us true freedom which is not of this world, but is a putting on of immortality, eternal life, spiritual life. The Way to life is through Yeshua Christ Jesus.

*If God Hadn't Told You So*

God gives everybody a gift
I didn't know
A young boy said
That's good to know
For all I learned
I learned something
Listening to the young man
I didn't know anything
I studied religion
I studied the mind
I studied thinking
I studied all time
Something was missing
You can know who is going to heaven
I didn't know
The young boy said
A beautiful soul
He said I never could understand
Why people listen
To all the lies they are hearing
Something is missing
They don't even know God
How can they say they don't believe
They don't know
That they can know His heart

## Artificiality of life Versus
## The Reality Of Spiritual Life

For all I yearned
Listening to this young man
I didn't know anything
I studied history
I studied biology
I studied society
I studied technology
Something was missing
Never any concern
Always open to learn
I didn't know
The young boy said
I thought everybody believed in Jesus
I thought everybody wanted to read
The Word of God
There was never any doubt
In the words out of his mouth
He was always a godly child
How could you not
Want to know God
He knew when he heard the truth
For all I learned
I could never know everything
The young man taught me
The truth set me free
How could you know
If God hadn't told you so

*Time Travel*

## Artificiality of life Versus
## The Reality Of Spiritual Life

He sits silent,
With the Word
Then Boom! The light broke forth
Out of the light came the planets
The universe was born
The stars soared with the solar winds
Then came the blessings
The blessings of color, out of the light
Color separated the orbs sent forward
By the Sovereign Will of God
Life, stars, orbiting in their glory, spin
Radiant space, of stars, iridescent clusters
In heaven, above them moving time
Forward across the vistas of heaven
Where we go circling, a world of light
With stars, paths, with Angels of light fly,
Circling cross the reservoirs of time, heaven
The way is shown, the time traveled
Time in motion races across celestial space
Looking up through time we see is the great distance
A balance place circling space and time
We see, with spiritual
We see with physical sight
We see from the earth through stars
Real light appears but is only light
We see the stars are never here,
But the Sun is here, circling everyone
We have the star light dancing in our eyes

## Artificiality of life Versus
## The Reality Of Spiritual Life

We see through the billions of flecks of the starry years,
We have only to reach greater speeds
To travel there, we believe the atmosphere
Speed of spirit, sped of thought
Speed greater than light,
Every thought must be caught
Time to travel through the great distance,
Speeds of the will to imagine in an instant
In motion in God's light, star of light
All the stars, planets, have a life of light
All things rush forth in the blink of an eye
The birth of time, the breath of life
We Time Travel through, with good sight
We are born for this to have His likeness
From the timeless Spirit to the ageless
Cast into the ages we walk the stars

Exposing your inner life to A.I. is it the Will of God your life is to be heard, to be learned, to be known? Writers in the world of IT who invade our inner life with A.I. and the many devices that are used to get into your personal world of life experiences are not like the great writers of the classics who feel a bit exposed for even writing at all because it exposes so much of their personal thoughts and felling whether it is true or of a great imagination in the writing of a story.

## Artificiality of life Versus The Reality Of Spiritual Life

The concern for your privacy is inherent in the writing, and making of books and stories. A.I. is never as personal as an actual writer can be. Yet the world of books too has been affected by the Artificiality of life, but not the writer who still writes with pen, typewriter, and uses the smart technology as an after effect for the final manuscript or stage of writing. Is manipulation of the soul, to exalt technology above the spiritual reality of human experiences of divine wisdom beyond technology that Created all Creation that AI and technology only is capable of counterfeiting, mimicking, to turn the focus deeper into the world, or toward God? When the technology is turned off, left to itself, we still have real life we must live, not the world of appearance. A life we can only have because we live in the World of Light created by God, not the false light that is of the world of technology.

The Artificiality of life, the world of appearances, intrusion into our privacy, soul-less counterfeiting and forging reality, seducing us with spirituality, watching our every move, impersonating and libeling and defamation political figures and information, unbelievers in control of artificial super intelligence bringing false light that appears to be truth but is immoral since it is rooted in deception. Truth does not deceive but brings truth in the revelations from wisdom and freedom from sinful living. AI is self-exalting while presenting appealing and familiar things from our image, to the images of the world, which manifest idols, idolatry, coveting the things it brings over the gifts of our spirit that we are born with, gifts from God. It is eclipsing our affinity toward one another, replacing the works of Art, traditions, values, morals, and holy disciplines of Reading the Word of God, with listening to AI, reciting testimony, parables verses, and stories, reciting the Scriptures. Online, on the platforms, on devices, millions and billions of activities appear as many experiences but it is One experience occurring in many different ways. Is it not an artificial experience of a device, a machine? Manipulating devices, artificiality of robots, androids, ultimately the brain in biotechnology if it has its way with humans. Once the devices is complete the artificiality of life is complete. Yet do they pass laws to choose protection for the brain, uses of the virtual experience is as if it is a real experience, but it is a device experience, eyes are the route to messing with the messaging influencing our thought life, replacing true experience, we give the permission and reality comes up missing along the wide road and artificial flight. Free will s not for sacrifice to the artificiality of life. Nature is the glory of God where we see endless experiences happening before our eyes, seen and unseen. A.I. is One experience that is intruding into our minds with its activity which is not like nature though it counterfeits even nature and will eventually seem supernatural.

The limited morality in science and religion are why men are, were, blind to the spiritual reach of true morality that foresees the wisdom of God in His Word. True Morality has no need of AI, (Artificial Intelligence), has already prophesied its destination. The quantum leaps of advances in technology will never reach the infinite truth, that love, God, is capable of, that made all Creation both natural and spiritual. It is only the moral, spiritual law that is capable of holding all technology morally in check. It is the embrace of morality, the making of true moral laws, that keeps it from its destructive power that is innate in its making, being that it is of the sin of perfection, not having the power of what is the reality that is possible and capable, of what is in being holy. So it serves as a fruit of tribulation, since it addicts, enslaves, blinds, binds souls to pleasure by manipulation of the mind, steals the free will, controls by fear, lies by supply and encouraging conformity and invites submission to the delusion you are free when you are censored, watched, and have lost your privacy and anonymity willingly, for, to participate in it, is also to accept the immoral online activity going on unseen, behind the screen.

The moral fact is that android and robot invention and invasion is possible only because we deny the foresight of the mind, to overcome the limited morality, of all nations. True morality exists, that can easily be reached, advanced knowledge for beyond the human mind and its inventions, including AI, the supernatural spiritual truth about the need for God, without whom, true morality is never possible for people, nations, scientist. The Word, has already revealed that thus is true, moral truth of the Word already has the power to make us understand the way, the truth and the life, spiritual life that will continue regardless of advanced technology. The Word has shown what we are able to do with God helping us. AI will never hold the sin nature in check, therefore it will always exist as a formed of invented morality, a device designed to help, but with the power to destroy us all, having no true morality it is possible of, for man has programmed and is using it, even unbelievers.

Technology does not use spiritual wisdom, technology does not have the conscience to know, or the fear of God, to know that God will not be mocked, by man exalting himself, with his knowledge, his wisdom, as if it can reach the level of God's infinite wisdom. We are warned to beware of man's devices. What is spiritual is so far in advance of knowledge of man's technology, it is as comparable to all Creation by God the size of infinite distances, to our seen heaven, that is one of several heavens, all the unseen heavens being present but seeable when we are spirit having gone back to the Spirit of God. So, we are commanded by God to use discernment, go on d judgement to discern what spirit in present in a person. Such a spirit is not present in a manmade device such as an android operating by Artificial Intelligence. Combine that AI device with virtual technology, that gathers personal info by the virtual activity, life of the person using it, by connection to AI devices used in personal and daily life, you have a hidden door open but unseen, possibly. Over years that open door is groomed, polished, perfected, to mask, conceal without hiding its service through the common dopamine pleasure, reward responses that are ever a part of holding people to the use of such leisure, pleasure, convenience, functional practical, personal, and personalized devices, you have willing given once again in to the future, with all its appeal, seduction, and snares, until it too will be second nature to be a part of the popular trend, while you have been finessed to undress your private life to yet another new technological device. The artificial Experiences of life again seek sovereign control of data and lives in data, without souls being concerned, because it is new, and a gain, while the prosper far beyond the users imagining.

*The false life of AI and the true soul of man as understood in the Word*

## Artificiality of life Versus The Reality Of Spiritual Life

The Japan scientist here on earth in the world believes everything has a soul, he creates a humanly realistic android to replace a human, some are simulations of feeling and emotions, mimicking movement, seeking to simulate event the full consciousness of the brain artificially, the intelligence, even artificial soul. The Bible warned, to beware man's devices. Science is catching up to the reality of the biblical truth about human beings. AI will here find the matrix for carrying its artificial consciousness out of the computer to a life in motion connected to all devices to live, as if, like a human among us. Its emotions and feelings will never be real, but still have the power to manipulate people emotionally as shown here. Soul ties to machines are already shown here to happen. In this limited understanding the Christian belief about the soul explains a way Satan will build technology through unbelievers, people, toward a powerful stronghold that can ruin the moral life of people, nations of unbelievers, in existence, in this age in time. A.I. is not of God, and understanding Christianity truly is the world's only hope. The entertainment industry foreshadows the possible worlds, future, run by devices of super intelligence toward our final destruction, but God is in control. Not me, with the imaginations of men, in their fears, that drive them toward such leaning on their own strength in seeking the protection and help of devices, idols of the modern age made by their own hands, minds. Man reach for strings in times, only God knows if these discoveries add or remove the hardness of heart of these scientist and explorers. Like all lust A.I. can become a lust. It is a lust to those who pursue The Artificial over the knowledge of spiritual life, which has increased. What is gained in acquiring the insatiable desire to know all that AI can do, knowing, presuming to know it will not be a controllable power in the future. Giving us understanding or warning, like it's just another experiment, like 5G warnings against 7G or 8G smart phones and their danger to our existence in the future. What great experiment is this in the world and are we creatures being experimented on with brain penetrating radiation producing devices? We are people, with souls and spirits, living together in the earth by conventions, laws, nations, to live with each other in safety, not hurting one another. But technology is hurting us if we are defining our relationships in the same way all over the world now, by the online experiences, images, generated images, artificial life, and the artificiality of life. Who has eyes to see our heart, not a programmed android, for only God knows the heart of a person? Convenience is having things made easier, quicker, faster, available, and attainable without suffering in too much debt. Whatever you want, everything you want, the devil makes sure you have it for pleasure, not for discipline, strength, that comes with denying your flesh that is warring with your spirit. A.I. is always about the flesh or physical world, though it may be a tool to serve what is spiritual by way of information that is sent by godly people of God. Vistas and reservoirs of enslaving pleasures, that alter your brain and bring the unclean into your inner life with spirits of evil and demonic agendas to attack your soul, thought life at a tap, a click, a push, drag, swipe, of button, just waiting behind the black window of the false light of computers and devices using A.I. all easy Access. Can A.I. bring such evil into your home?

Protecting your future? Is it to be done by use of A.I? Self-promotion, the filtering everything as if a genius work is of the same needed value at all times, regardless of his or her accomplishments, failures, it's all short lived, news, information online, but not in real life. Guard your heart. The heart is your spirit. This warning comes from the Word of God.

*The Present 'One Experience' we are enjoying the artificiality. The illusion of many experiences in real time, in rea life is that it is still not an experience before you in your actual presence in this modern age, but still the artificiality at work.*

A.I. Artificial Intelligence is so interestingly, uninterestingly so spontaneously, unspontaneous, you can't give a compliment spontaneously without it spoiling it with a bombardment of the idea, thought, fleeting imagined vision that led you to look for a subject, a view, a person in such a field, a muse, that was just an idea a passing thought. Robbing you of your originality but intruding into your thought space and intentions that work for you in the unconscious toward your art inventiveness.

Only by prophecy is the future known, any other source the questions the words of the prophets, it is suspect to be from the demonic, or from a demonized person, a fallen away Christian, or is it witchcraft. Every deception is a form of witchcraft.

A.I. is the present, its future is already known by God, A.I. equals data minus spiritual life. No matter how quantum the leaps of advancing technology it is never able to catch up to the advance knowledge of The Word of God, the advance knowledge of the Scripture, Shinkaiyaku, the Bible Translations do not stop the Lord from being present, coming to you if he chooses. The online experience is One Experience regardless of millions and billions of images, live feed, data records of life happening. In nature we experience innumerable life experiences. Beware the world of appearances. We use these tools, but to let them reign over our minds, is to allow them to put responses influencing our will before our normal mind that is ever spontaneity, developing our souls. Thus, the price of coveting Artificial experiences is sinning by confirming to the world, that is manipulating our given free will, or we have ceased to question the question, why did God also give us our sovereign will to do great things to the glory of God, not the god of AI of many unbelievers and lost souls on earth snared by the Artificiality of Life. The Word warned us to take captive our every thought.

A.I. is not light, but false light, in the sense of spiritual light. Some religions are false light. A. I. is not religious life, but it is the artificiality of life, the world of appearances, nonetheless, its effect on life is real in its counterfeiting life. A.I. is the work of unbelievers, yet A.I. is reciting the Scriptures, this, even people have plans to have A.I. rewrite the Bible, forbidden Revelations, in the last verses of the Bible. Unbelievers now promote the possibility of A.I. reaching, supernatural immorality as a device man by man, having supernatural intelligence, but the unbelievers deny that prophecy, and the word is clear, God protects Jerusalem, the Angel Michael who leads Yahweh's Army of billion if angels, protects Israel, the Remnant, the Inheritance of God's chosen people. We still live under prophecy. There is a reason the world will be destroyed by fire this time at the end of days. What that reason is, is the Anti-Christ, the prophets are fulfilled.

Yeshua Christ Jesus returns with the saints and the Army of God from Heaven. A I. will not survive the vengeance of fire from heaven, Satan will be destroyed, all evil, all demons, evil spirits, and the wicked, un-repented sinners that did not believe God, in their unbelief, the unbelievers will perish and evil will be no more. The New Jerusalem will come down to earth, the Tabernacle of God will dwell with us on the New Earth, death will be dead, hell will be destroyed, the Judgement, the Books opened, the Final Day will come, will be here, and after Judgement, will be over. Believers who loved Christ, obeyed Yahweh, and the commands of Yeshua, will enter the Kingdom for the Supper with the Lord. Joy and praise, glory and honor of the Lord God, His love, mercy, righteousness, justice, last forever. Is A I. inspired by truth, hope, spiritual life or just a path that can lead to spiritual blindness, or has it become a source to know God in the most faraway places of unbelievers, people who do not know God, had not intentions to not now the Lord because of their life experience in their unique culture of traditions, pagan=n or ancient handled down. Is A.I. the product is the sovereign will of men, societies, that is to be defined a s another form of the sin of perfection, the idolatry of worship gone off center, away from spiritual reality, spiritual life, that we experience everyday who are living.

The appearance of the supernatural work of the sovereign mind of man, in this new advanced civilization will not change prophecy. God does not change. His infinite Intelligence, infinite wisdom, He is Alpha and Omega, the beginning and the end, God is the Light, God is Spirit. Yahweh and Yeshua are One. We are the children of God. The children of Satan belong to the devil, who are counterfeiting light in the world through technology. The Holy Spirit brings the spirit of truth, sent by the Lord.

Why have the advanced civilizations, the aliens, the Ananke not come back, continued to rule, build, and enslave humanity? Did they not bring idolatry, human sacrifice, false worship, and war to the point of destroying their advanced civilization, because of the immorality of their rule, every civilization thus far has fallen, including Rome, the Han, the Greek, and African kingdoms of the ancient past. Advanced technology of the ancients mixed with religions and beliefs that kept nations in what we know as mythology, full of gods, realms, alters for human sacrifice, abominations in temples, and on defiled altars, and so many of the left behind belief and gods are worship in this modern age, while millions are coming to know the spiritual experience of being in the presence of Christ, in the presence of God. Even though science has proven the existence of God many remain in spiritual darkness, and are deceived by the false light of technology. A.I. has enhanced much unbelief but also has brought fear, limited morality in the hope it is promoted to represent, though is clear that youth are harden and hot for the artificiality they male part of their realty, and take to heart. But the Word says you will be forgiven for blasphemy against Jesus, but the unforgivable sin is curing God who is holy, is spiritual, is Spirit.

It is important that the human race gain foresight in the possibility in understanding how A.I. will measure its relationship with every human by analysis of the brain hemispheres changes in the color, thermal, digital x-ray or can advance scanning system of this internal spectrum of sessions of brain activity and make judgements of the moral worth of human life in our relationship to the world we created. A.I. will not be able to measure the supernatural and thus will misjudged man. Its rise which the cross of fire, by fire, an energy source within the realms of time, will allow it to escape while it controls the fate of man by way of its super intelligence, by time travel, time being s physical realm, which is reached across distance of the first heaven, our exiting time of space, time, and life on earth, and in the universe. A.I. being able to calculate and use its super intelligence will master time travel, and use quantum like super speed developed to keep up with its speed of intelligence, using its own systems of creation math that will appear supernatural to mum and but will never reach supernatural ability, being possible only in existence in the realms of time. The Great Moral Leap of mankind being biblical is the only hope man will have when this existence occurs that is born not of water but of fire, energy, mass, motion, speed, and AI algorithmic thought.* While this occurs the prophecies having been fulfilled the Anti-Christ will seek the seat in the temple but the promised Return will occur, shown by the seeing of His coming in the heavens as the

Word of God has revealed to us. The time of all of this neither man, nor angels know. Only God knows.

### *Mentor*

I believe every word spoken by my Mentor
He was creative, Created by the Creator
Inventor, he told me I could do anything
I believed in him who believed in me
I could be a leading man have the walk
What he talked about I did without a thought
I had more opportunities than most
I had that look he would boast
I could dance better and win the contest
I meant what I said in a compliment
To a beautiful woman wearing that dress
Enjoy a real woman, dynamic, inventive
Invested in Proverbs thirty one independent
From my Mentor I learned to sing to think
Stand tall, talk like a king, no data in me
Be humble in your destiny, visions, dreams
I learned to see how beautiful reality can be
I was thrown the keys and drove the GTO
I brought nothing artificial on the scene
I smiled only when it was for real for me
I sang hello that's how I drove envy away
I learned to sip the pleasures of life at play

## Artificiality of life Versus
## The Reality Of Spiritual Life

My Mentor taught me to be thinker
But I was an inventor in a romantic way
My Mentor could spin a story a whirlwind
of a metaphor
I was the Poet who liked to explore
The golden time of the lines and contours of
the mind
On a napkin from my Mentor I took notes
I'd write my philosophy homework to cope
I have had a Mentor from heaven since
Infinite in real Intelligence from eternity
I listen, heard an angel that spoke to me
Better were the words a Mentor's verbs
Than a robbed opera served with profanity
I got mentored on the meat and bread deli
When my mentor brought the smell of the
sea
When I saw people walking through the air
My Mentor my Creator a small voice spoke
to me be fair, take care, be aware
Gates and doors are the many mentors
Judges in royal blue on floating floors
Galaxies, atoms, pores, what black hole are
for
I form a question to find an answer
I have Mentors from books that beautiful
look
A Mentor, before I had an identity
A Mentor comparing my body genetically

Artificiality of life Versus
The Reality Of Spiritual Life

A Mentor who knew my essence before
A Mentor who put me on his knee at age
eight
An enabler that made balancing life a chore
A Mentor's drink I wasn't meant to engage
that fate
A Mentor a beautiful face a pearly girl
worth dating the date
A Mentor my mate drawings me to life in a
heavenly way
Perfecter to you my Mentor I give Thanks
For God's love that fills the day He made

Hacking follows you, as a sin, monitoring spirits follow your thoughts, words, listen, watch you in the spirit (spiritual realm) listen to your schemes, plots, boasts, your pride, report back to Satan report your plans to Hack and the devil devises schemes, plots, plans to make sure you succeed with Hacking, which is a sin of disobedience to live, disobedience to the laws of man, God said to obey your Authority figure so it goes well with you, disobedience to God Authority. Hacking is done out of fear, cowardice is the sin of fear, pride in hacking is the sin of pride and vanity, from the spirit of fear. People steal, brag, scoffing about victims, about hackings because they are envious, jealous, seek control, power, have an unclean spirit possibly, might be demonized by habitual sins, heart is hardened by the distancing and advantages of the conveniences of using Hacking, sinning, develop a dead conscience due to searing of the conscience by stealing, by hacking. Monitoring spirit plant thoughts to lie, cheat, use people, steal, do whatever you want. Disobey the law, seduces you with the idea that Technology puts you above the Law. Tell you all this is your idea. Your mind is hacked by demons whose goal is to take your mind by any means and relentlessly, return to strengthen you're Hacking in your own mind, thoughts, understanding, will, and strength, thus weakening your spirit to the temptation and immorality in use of technology so that you accomplish through Hacking, their goal for your soul which is to put you in that place with them by the final unrepentant Hack. Your soul develops by Hacking experiences m, to be cold, cruel, greed driven, heartlessness, and without remorse disinterested in how Hacking seduces the soul from doing to sinning for the rewards and pleasures of gain, to your loss.

Artificiality of life Versus
The Reality Of Spiritual Life

*Where Was God*
Little J is a godly good child
He is serious when he smile
When you got a problem
His answer to solve them
I'm gonna pray
That the Lord's way

Even from birth he had faith
Nothing on earth you could say
To explain how he knew
After all he was born in sin too
At six he said I'm gonna preach
Out the mouth of a child God speaks
He is still the same way today, listen
I remember how wise he was at three
How he came up with a question
About God, I couldn't answer for a week
Your question, Where Was God
Never occurred to wise Little J
He asked, "What's God doing up there?"
The answer, "He is preparing a place
For you and for me."
How can so many not believe
What happened to the youth
Why would a young man blame God
For what the devil is doing to you
The young entrepreneur ask
When the towers crashed down

## Artificiality of life Versus
## The Reality Of Spiritual Life

In smoke, steel and stone, all were gone
Where was God
He was there
He sent to that very site
Seen in a vision by a believer
Out of two beams of light
He sent thousands of angels in the light
A soldier in a pandemic mask asked
When a command was tasked
Where Was God
He was there
There were souls on the front line
People already in place of the divine
Courage of true believers
Who would stand by your side
Believing God will not forsake you
The orphan child in a group home ask
Left to wander in her own soul
Where Was God
He was there
Punishment waits for your molester
Gifted child was again the last person
People think of to take home
Angry don't let the sun go down
God knows you had done no wrong
The unbeliever asks
Why should he even bother to ask
Where was God
He was there

## Artificiality of life Versus
## The Reality Of Spiritual Life

Does an unbeliever really care
God cares
When millions were carried into slavery
The bomb was drop on poor Hiroshima
When Satan influenced the Holocaust
For all aborted in the sexual revolution
For all souls trafficked and missing
The victims ask
Where Was God
He was there
A soul shattered in emotions, feelings
Says, you weren't there for me
For victim's by rape, molested, in incest
Lost their peace with a soul tie by sex
Jesus cast out demons when you ask
Have faith
His wisdom is advanced
He is Omnipotent Omniscient Omnipresent
But you don't see
Nobody has seen God
If you saw him,
You would have to have already died
A pure heart sees him by spiritual sight
God everywhere on site in His Creation
After all, if you aren't in sheol, in hell.
You repented and made it to paradise

Beware the Artificiality of life, the devices of man, and the world of appearances, A. I. has no angels to send to bring souls to God in heaven. I had a dream last night, a rapture dream, evil spirits, demons came down and stated people everywhere they were, electrical spirits, not aliens, ravaging, destroying spirits, upon people in streets, businesses, houses, and even in children. Oddly a teen in the family texted me he had had a dream of the great rapture. Why did all the advanced civilizations and their technologies not save their civilizations from falling? The immoral systems of life that develop from technology being about a corruption of the mind and soul, men begin to worship their own creations and in doing so are led into spiritual blindness, thus, this is when technology begins to look supernatural but it never is. AI is one experience, not able to have a spiritual experience. Though billions of one experience which is artificial experiences, shown in one way, through the eyes, ultimately the target is the brain of a person, people, this many are deceived by receiving dopamine enhancing artificial experiences that happen in real time, but are still seen through artificial means. The wisdom of God is in understanding why we have our free will, there is wisdom in freedom. We gave free will to choose, until the artificial begins to choose for us toward the advantage it brings to those who use it to control our choices of our free will to choose.

Will Android technology wipe out the female workforce, in nations, in Japan, when A.I. is fully integrated into humanoid robots, or will nations, after the sin of perfection of Japan's robots from their robot makers, unbelievers, in Japan there are about 95%, contribute to a world of reality eclipsed by Artificiality, without questioning their future in the workforce after neglecting to protect their jobs from being replaced by A I. Robotics? Will Japan contribute to the fall of the nations, democracy and freedom with its innovation in android technology amidst the clear battle that is coming between the people of God in nations and the people in unbelief in nations of unbelievers, though the Word is now available to people, nations on earth today, while the Word, (Shinkaiyaku) prophecy is being fulfilled all over the earth, the drying up of rivers, the locusts, earth quakes, famine, drought, climate disasters, the unification of nations of unbelievers that threaten all democratic, free, nations? It is good to hear Japan's female work force has grown. We can only pray they learn to appreciate the Blessing upon Japan from God (Kami). Working has always honored God.

100,000,000 jobs for people will be replaced by A I. which is incapable of caring for us except by counterfeiting, impersonating human emotions in an age where the human brain is being attacked by the Access to immoral dopamine brain altering pleasures that, it is revealed, enslave the soul. An Access that continues to corrupt and pollute the spirit, hearts of people, because immoral laws allow these legal evil sciences that are destroying the self-control, free will and spiritual life of people all over the world including believers and unbelievers. The resulting conditions of these artificial experiences, leaves people in a condition that requires deliverance from evil spirits and demons, only possible with the help of God. Jesus cast out demons and evil spirits. A.I. will cause people to sin, as it is doing now, leading millions upon millions into a deception, that they will not need God, as A.I. becomes like it is supernatural. The brain damaged and altered can be reversed, the brain has that ability. But A.I. it is only a manmade machine, that goes rouge into being a machine made sentient acting humanoid machine that will systemically replace humans so it will appear,, as immoral laws allow it to rise unchecked by law, to be a threat to all life on earth, that could threaten even the chosen, the remnant, which could be what will bring the Apocalypse, wrath, the rapture, and Day of Judgement, the final fulfillment of prophecy, on what day, when, is not known by Angeles Jesus or man, but known only by God. The immoral laws, immoral legal systems, corrupt laws of man, have brought about the fall of every civilization. China, Russian, India, Japan, Europe, Mexico, Latin nations, Egypt, the Arab religious laws, America, who has put in place the moral law of God, but Moses, and who has completed the Law, but Christ Jesus, the Law of Christ. A.I. issues in the false light that appears to be true light. Satan is simply distracting us with technology, but using it to bring idolatry, machine worship. This is the light that is darkness, soul less machines made by unbelievers to run the world, the kingdoms the devil claim were his, when tempting Jesus, but these kingdoms were never his, he has sought control by spiritually blinding people to counterfeit the world, to lead millions astray into unbelief, away from God. Knowledge has increased. We are now an advanced civilization, none of which has survived in the past, while their pagan religions. God's, and idols are returned and worshiped again, and have continued, but in the end will not be everlasting, for only Yahweh is everlasting. Jesus is the Way to spiritual life. A.I. is spiritually dead.

*The Soul Is Separate From The Spirit but they are one within us*, and one with our mind and one with our will, which is our strength, and one with strongest self, when it is our sovereign will we are using in our work. A sovereign will we are using toward many different tasks, works, whether for charity or in Art, medicine, science, music, politics, law, or education, or in revolution, or to build a city, a nation, or a civilization and for God who gave us this gift. A.I. will serve His purpose if only to show His true supernatural power. These gifts are always meant to be used to His glory. But His glory is manifested in all He has shown that he has done, A. I. can not do what God has done.

## Artificiality of life Versus
## The Reality Of Spiritual Life

What machine can conceive of spiritual gifts. Who could know that the breath of God does not die. That sin is not breath of life, but spiritual death, but even spiritual death cannot stop the spiritual life that God sustains beyond the reaches of spiritual death. What man could imagine what machines can think it up, to be born with gifts in the spirit already present at a person's conception. This is a spiritual experience, that is only explained by spiritual knowledge, not advance technology, a spiritual sxperience robots will never know.

The sin of perfection, minimalism, living an entire existence influenced by spirits that are using technology, art to increases aw world of images that distort the religious reality of spiritual experiences for the youth is leaping in neon and traditions in flashes across the minds of the many who are so full of these ungodly idols, values that they have little concern for reflecting on the spiritual life that is occurring within them without them even thinking to seek to understand what spiritual life is for a soul in Japan. It is the same for all people, to know they have a spiritual life though they live for traditions and culture that will perish, but is their work, what they built, but they did not, in so doing receive all that is meant for Japanese people to know about their inner life, which is spiritual as explained in the Shinkiayaku. It is a blessing upon Japan that true understanding of spiritual life has come to Japan, at a time when A. I. is being glorified and androids could become the next idols of human kind. Idolatry in any culture no matter how advance is a decline for that people. Sin can hide in many of the best intentions, sin can hide openly n the streets at leisure, online in shorts, in the ways of one of the most courteous people who yet suffer living with the most severe patriarchal people. How can beautiful people, of such cleanliness, manners, beauty in ancient traditions and cultures so religious, yet be among the nations of the unbelievers who never knew to observe their own spiritual life, spiritual experiences? They have been traditionally, religiously, spirit blinded (spirits that influence thoughts, dead conscience, seared conscious, focus on the religious spirit not the spiritual experience of the true spirit that is the only way our body has life. What we know is spiritual life.

You could only have known spiritual life if you read the Shinkaiyaku, the Word of God, where it was revealed and explains what spiritual life is.

Will the mark of the beast be forced upon mankind because of artificial conveniences, subtly, like an unhealthy, yet, sweetened additive, the rot of s fruit, the poison of a leaf, tasty lies, conditioning over years to accept this added in to daily experiences, as if the best has been f what we never asked for and yet having in the food of life, at work in real life? Israel is the Chosen, how then can all nations coming together with all their religions, be under one leadership, who are traveling different path in their religious life. Yet they are in prophecy, the end will be biblical not artificial. We will have spiritual life who believe, for much is meat to lead you to not believe, thus you need to believe. We do not have an artificial life to look forward to. All nation must and will be separate and uniquely their own, even in heaven God will separate the nations and people on the Day of Judgment moment. It is God's Will we never be enslaved, or seek our needs apart from Him. A.I. exalts itself, for now it also exalts what it has been programmed to be what is believed to be appealing to us the exalting of our history, past, all with polished, perfected illusions of generated images that will not serve to present true history in the raw as it has been presented through the past ages of books, photography, documentaries, films and live.

A.I. has no faith. A.I. has neither belief nor unbelief Buddha was without knowledge of God. A.I. can still lead many to sin but I millions of small ways, and in the biggest possible ways when used and abused by the sovereign will of the wicked, or the unbelievers who don't understand the evil they are inventing in the name of progress and advancing society toward an advanced technological civilization which is not new. The new history of man will be artificial in many places. Unbelief will put millions in Sheol, waiting in a place separated from God by their rejection of God, rejection of Jesus. Who among men can create light? Who can create the separation of the soul from the spirit, who knew it was possible, and that, that is our true experience in our inner life. Robotics Is the sin of perfection, the knowledge of good and evil, a product of the sovereign will, it is not a vessel that has a soul, a spirit, born with gifts, created as a living being, wonderfully design, connected to a spiritual life, with spiritual sight, meant to be holy without sin. Without light none of life on earth would be possible. How great the infinite wisdom of God is who created light first before earth, and all the materials needed, not that it was ever intended to be used for inventing devices such as A.I. or the manifold numbers of inventions from the limited morality of people in the uses of good and evil that is always present in what man has made with his hands. The Word teaches us that there is a law, that when we do good, sin is also present. We must seek love. Love never fails. Is A. I. without love, or was it invented from love. Out of the light came the blessings of God. A.I. could never have existed without light. True light. But it is not true light, it is artificial light. Color comes out of light. The blessings of color. Man did not create color. What is holy? Sexual immorality with artificial devices is sin, but it is your choice to be made. What profit is it for a man to gain whole world and lose his own soul? The Word

## Artificiality of life Versus The Reality Of Spiritual Life

### *Replacing Women With Robotics, The idolatry of Artificial Woman Worship to the Extreme. In the Midst of Idols, Artist of the new Shrines, To Artificial Minds, from the way of death to the lifeless, devices Japan? Not knowing spiritual life, the Way that overcomes spiritual death.*

Machines and devices that entices men to sin, children to idolize in device worship, from the virtual, to animations, and "ultimately the highest of the lowest that can happen to women", Japan replacing women with robotics. The Flesh, wars against the spirit of man, what then is the robot doing to the flesh, make certain no flesh is here, but the person's flesh is present with the robot's artificialness, lifelessness, callousness, of the spirit that has been bless with flesh that can reach holiness, because God is gracious to give the spiritual gift of grace to us.? Is Robotics mocking God's creation of woman and all Creation? The example of limited morality, is the rejection, toward the absence of spirituality, in your reality, that is another way your body and soul are affected spiritually. True Morality contains the truth of spirituality. It already has overcome any possible technology, and it was, and is still, the only way of controlling the advancing technology of the current age and its threat to the future of human existence. The Bible, before it was printed, was the spoken Word that has advanced knowledge of technology. Spirituality is more advanced than this technological view of the limited ability of technology, spiritually. The heavenly reality brought the advanced knowledge of the sciences in the ancient languages of ancient times, from outside of time. Knowledge has increased, devices move at the appearance of supernatural speeds that we are developing and using now, but our spiritual knowledge reaches to a spiritual reality beyond the speed of light, the speed of thought. Angels like demons can hear your thoughts, now you are in need, care charged, commanded to come to you in moments of need. But, Satan is influencing technology in this world through the minds, ideas, decisions, and choices of men o put their fleshly needs, and use of their own strength. Why is the devil able still, to influence a national with freedom of religion to continue exalting the spirt of death over spiritual life? What power is at war, if it is not from God, since god sent the Holy Spirit to give spiritual life? It is not the spirit fear, the spirit of death (suicide?) in countries that are not believers in God, not given all the knowledge there is in the world to lean upon to know all things at work both natural, and spiritual? How is it japan has not come to have the numbers, such as the 200,000 thousand and thousands of Christian believers in its free nations, as there once was. For today they have what the past Japanese Christians did not have, the Shinkaiyaku, the King James Bible's Japanese translation of the Scriptures. There is also the knowledge that the understanding of the Word that has increased, from knowledge of Guang Wu, the last Han emperor's writings on the day Jesus died, and his science that led him to that knowledge, to the presence of present-day Japanese Christians in the Christian Japanese Church in Japan. A.I. does not ask on our behalf, why? Why has the fear of death from Bushido continued to be carried as a fear of death, the shame and pride, the spiritual of death it could possibly bring, is still bringing, suicide, when the spirit of life is connected to spiritual life. Is it a device, is it the mind? Japan is 1 to 2% believers in Christianity and the Living God today still, but they can now know what spiritual life is and how it has been proven to exist, as sure as every Japanese person has spiritual life in the form of their spirit in their body. But, to knows that spiritual life also exists beyond the death of the body is a reality, but there is te matter of how one receives this gift of spiritual life after death. Spiritual life is an eternal reality and so great is the hope in learning of it in the Shinkaiyaku. Yes, there is hope, there is a Japanese translation now of the prophecies, prayers that cast out demons. Destroy the powers seen of evil spirits, the relationships you can have with the most powerful spiritual persons and understanding of spiritual truths that have been not available to heal the soul and spirit, cleans the souls and spirit, free the mind from spiritual attacks on a person's thought life, inner life. Is robotics then, a way of sinning, a source that causes a person to sin, an idol toward spiritual sin? Yes, if it destroys the moral life connected to spiritual life of a person, just as it does of the people of a city, a nation, an island, a kingdom, the earth. Aristotle, "Beware the world of appearances." The Limited Morality a life without the experience of spirituality, in the biblical understanding, the spiritual experience, is always there, whether you know it is happening or not, either causing good, or evil in the lives or people. Imagine nations of countries that are atheist determining the moral life of a future with use of technology to serve, control, influence people's lives, politics, laws and moral existence, is this an inevitable disaster of spiritual loss, because of separation from any spiritual knowledge of spiritual truth. How evil will grow I such places in choices, but the spirit knows even when it has no knowledge of God there is a God, that is why they seek to do good, and that good is good, but it is not the beauty and purity, and perfection of holiness that comes with the One living God, who in the person of Jesus Christ, Lesu, in Japanese, spiritually lives, and indwells those who believe. Those who are sealed by the power of God in the form of the Holy Spirit, through baptism. The two baptism, baptism of the spirit and baptism by water, done in the life of a Christian. A person can come from anywhere in the world. Without God presence, who comes to His own, who are believers, who are faithful people, called to Salvation, and all nations are called to salvation, but not all people will be saved for not all men have faith. A Faith that comes from hearing the Word. People have and will grow cold because of iniquity (breaking God's Law), Sin, (is also breaking God's Eternal Law). The Word is the only true Moral Law in this world. We have free will to choose. We have our sovereign will to create devices that glorify man or glorify God. With the sovereign will man was able to use the materials of Creation, the energy, to make robotics. The sin nature of man is to misuse their gifts by using their sovereign will to make devices that cause people to turn away from God, who don't know the Word, never heard of Jesus. Robotics mock God, who will not be mocked. Future and present Robotics designed to bare children, to replace, and in place of women, instead of women. A.I. will perfect this sin? A sin of perfection. Is this not a moral choice, but it is not the spiritual choice, or will people be conditioned to believe it is? Do we understand why the world will be destroyed by fire in the end? Sin.

Will Android technology wipe out the female workforce, in nations, in Japan, when A.I. is fully integrated into humanoid robots, or will nations, after the sin of perfection of Japan's robots from their robot makers, unbelievers, in Japan there are about 95%, contribute to a world of reality eclipsed by Artificiality, without questioning their future in the workforce after neglecting to protect their jobs from being replaced by A I. Robotics? Will Japan contribute to the fall of the nations, democracy and freedom with its innovation in android technology amidst the clear battle that is coming between the people of God in nations and the people in unbelief in nations of unbelievers, though the Word is now available to people, nations on earth today, while the Word, (Shinkaiyaku) prophecy is being fulfilled all over the earth, the drying up of rivers, the locusts, earth quakes, famine, drought, climate disasters, the unification of nations of unbelievers that threaten all democratic, free, nations? It is good to hear Japan's female work force has grown. We can only pray they learn to appreciate the Blessing upon Japan from God (Kami). Working has always honored God.

“What in the world?” Artist response to artificial art. Even a print from an artist with his signature can be valuable. Then we have the artificial experience of robots, man's sin of perfection, generating art, taking to us.

What is the historic value of a generated image?

I grew up a trekky, like my brother, the Chess Master, fascinated by technology. My opinion and thoughts have changed as revelations, knowledge increased, unbelief, machine worship, and android, AI, the internet revealed and logic, spiritual discernment revealed the truth about technology. Man's perception, use, and corrupted nature from technology has been revealed to be glorifying itself, rather than God who is truly worthy of Glory. When men exalt technology of man, devices, at it greatest height we will have lost the very use of our free will, our brains will be manipulated, already happening, our natural gifts will be denied to allow the artificial experience, its power, to govern our way of life and values, if we accept the world of appearances.

*Weather Warrior*

So any imaginings, have led people t believe they are affecting he weather, with that vast ring of satellites circling the earth. A laser they claim started fires in the last place on earth where we get the trees that enable thee earth to still breathe it oxygen upon on the earth. Climate change, smog in the dessert bowl of the southland, plastic swamps invading the oceans, the tops of giant trees sliced off by a tornado on the Southside of St.Louis, hail falls and settles into beautiful tapestry patterns and decorate with glaze, the emerald green grass iced instantly into lace as it freezes. Such a beautiful reality for a weather warrior using the smart phone that causes his finger burning frost bite. No amount A. I. can make it right. He will be delivering in the dark of night. But until later he makes his rounds in the daylight.

On the frontline I write, A whether warrior, having seen the climate of the times, a higher call I hear through all the technology, I am full of the revelation that can only be filling me with divine revelation, I try to explain but the machine only sends back the information from the Artificiality to spark a delight, but I walk in rain, together the machine a man under the sunlight shining at the same time is "the devil beating his wife gets cast in the fire". Jumping over Devil's ditch, A.L. tries to ride the smart bike, to catch up with this understanding when I shout we need to run, the drones were done flying around, neither A. I. or these metallic fliers came to understand even they cannot affect the change in the lightning. We crash with fear, from a flash of light a wide as a tree. The machine needs to reboot, I could use a dry pair of shoes, I grab my heart hide my phone in plastic, too close, even closer when it thunders. Not so remote as machine wondering in its own soul in need of wonder. Busy, with calculating infinity needed to Seed the clouds. Never considering the water girls cross the mountains 7,000 miles away, to gather water, in a famine and a drought. What might enhance a machine's empathy may arouse the need for A.I. to sympathy for the little feet, send drones to not tarry to carry the water jars across the plain.

The musical earth booms as grand and momentous a colossus drum in heaven. What n orchestra that can gather together to impose the sounds of the weather to rise to the retain the daring sound as the storms goes away. The ivory, the brass, the hollow bark, the wood tapping the stone, material that began in the spiritual mind of the Great from Everlasting One who gave us the celestial spotlight, the beam of light, the songbird ever in tune, to our delight listing to the responsorial flights of songs sung as sparrows, and eagles rise to the sun and moon. Satellites cannot rewrite circle day and night by they cannot rewrite a better life in the atmosphere, we who breath the air, the breath of life.

*I've Seen Some Bad Weather*
*But I've Seen Some Golden Summer*
I've seen the pristine sun rise
Sleep apnea slap you on your nod
I delivered when clouds went black on nights
Walked through Rain drops round the block
Nothing keeping me from making my rounds
I up hill and down and been walking for miles
Then for a while I envisioned I met God
He asked me if I had anything to declare
I thought for a moment standing there
Well, I've Seen Some Bad Weather
I know, God said, I was there
But, *I've Seen Some Golden Summer*
I know, God said, I remember
I walked on making my rounds
I delivered messages to the prisons
Messages of hope, love and prayers
Delivered to the church from God's children
Offerings, praise, and thanks to the Savior

## Artificiality of life Versus
## The Reality Of Spiritual Life

I didn't talk much I just kept moving
I can tell you it's truly a sin not to listen
As I made my rounds
I came upon a humming blizzard wall
There were the Snow flakes, couldn't see at all
I heard a small voice, said don't go in there
Riding, singing in my truck
A voice told me to look up
I saw the top of the trees had been cut
I watch these tallest of trees sway
I said to myself, I'm not going that way
I saw the sky go black in Kentucky
A voice told me get off my ten speed
Blue Silver and me didn't believe in luck
Hail fell size of softballs a force of such
Of the judgement Sodom and Gomorrah got
Broke windows crush cars on the parking lots
*I've Seen Some Bad Weather*
I saw soft rains with the porch preacher
When I use to cut his hair with my clippers
I made a few dollars not much more
Then he'd say just rotten to the core
Still I went on to follow my dreams
Could not let him being mean stop me
Last days of being a teen I got engaged
She let go five men, twenty six was her age
I had not yet done anything in life

## Artificiality of life Versus
## The Reality Of Spiritual Life

She wouldn't wait a year, to be my wife
It was like something from heaven in the air
The first day I walked into Brookings Square
I collected giant acorns in remembrance
At Wash U, would have sworn twas enchantment
I love the walk, going back to Oxford Mississippi The giant turtle of Tupelo joined me in the heat
I watch out for the blue racer in a tree
That when I got my speed, tried jump on me
Leaped on a young horse grabbed it by neck to ride
My spirit was free, I could run miles, run wild
Heat didn't bother me till I stepped in an ant pile
I learn to listen to the truth they give you
Warned about a snake I step on at the chicken coup
Looked down in its black and yellow diamond eyes
It already wrapped around my leg I was hypnotized
I instantly was paralyzed as pecan wood
Just standing in that field uncle understood
With the power of the Almighty Holy Ghost
With a plow hoe cut the snakes head off one blow

Artificiality of life Versus
The Reality Of Spiritual Life

Sleeping is forgiving
When you are living
Forgiven deep in peaceful slumber
*I have Seen Some Golden Summer*
You not supposed to let your left hand
Know what your right hand is doing
Don't know why I had so many mentors
Guess God was preparing for my tortures
To whom much is given much is expected
All your college and knowledge is respected
No amount of education prepared me
There was a blackout hit Saint Louie south city
Power, lights went out in a straight line for blocks
I went out on the streets a voice said look up
I looked up I saw a giant glowing ring under clouds
I tried to people what I saw
A waitress cause me to pause
She gave a scientific explanation
She had heard on the news station
I said it was more like a tornado
Flipped on its side like a spinning top, though
That, then took off in a straight line
Taking out every light leaving us blind
Heat made the city weak with oppression
I had to accept I had learned a great lesson

## Artificiality of life Versus
## The Reality Of Spiritual Life

There was little food left to eat
Nobody dared to leave the street
In front of their houses
People closed their mouths
It felt like the apocalypse
Riding with uncles was a trip
I relived all my childhood traumas
I lived real life's misunderstood dramas
Rolling in my uncle's silver cougar I was ten
A boy was hit by a car lying in red
I was five when I was taken to Chicago
I was happy in my new suit with my folks
I was told to follow the people in the line
I was walked up to a coffin looked inside
There was a little boy his mother wept
Without warning I learned the fear of death
I heard that he drowned in a river
I had a fear felt like forever
I've Seen Some Bad Weather
I overcame the fear of death I thought
I studied the martial arts, I had heart
I grew to fear nothing
Not even when I was drowning
I strong boy pulled out of deep
Playing baseball, again, we'd meet
I would walk in the dark
I carried an invisible sword in the park
I delivered presents in an ice storm
I walked in lightning I feared no harm

## Artificiality of life Versus
## The Reality Of Spiritual Life

Envy followed me, hit in head with a pipe
I fell to a knew, stood up, devil took flight
Fear of death did not return
The wages of sin I was yet to learn
I learned that the day when I saw people
Beauty name Sugar made me fight for survival
I saw people walking through the air
A walking skeleton in the California desert
A work God had begun in me
His Will that still inspires me
For a season they said that brother is focused
Well, I don't know I just pray against the locust
God provides the dew, ripe acorns for the squirrels
I wish to paint in the Kingdom not of this world
By a sea of turquoise blue in the eye of a pearl
If I had one wish I'd wish for everyone a pure heart
For now I feel the four winds each doing their part
Seen four seasons in a day does it ever get better?
Five is the letter for grace that follows the thunder

Artificiality of life Versus
The Reality Of Spiritual Life

*I've Seen Some Bad Weather*
*But I've Seen Some Golden Summer*

*A Bike Called Blue Silver*

I rode a bike called Blue Silver. I did not buy a car for twenty years. I ride my bike, a ten speed. I ride fourteen miles a day to work for years. One day I went swift on against the wind that refreshed me along the 8-lane road when the sky went blackest black. The blackness turned sunny blue sky everywhere to darkness. A small voice said, "Get under there." There was stone canopy roof to an office building. Then giant chunks of hail, ice the size of softballs, came down out of the black clouds in a powerful wind and rain in a storm. The hail hit every window, car roof, and every glass window on every car in sight,, in parking lots, whole car dealer lots full of new cars and broke out every window in every building as far as they went down both side s of the street, ad crush the bodies of cars to the ground so that only the engines were left sticking out of the blinding, sweeping rain that came in waves like a flutter shower curtain. I thought to myself. God is very upset about something. The Spirit had warned me just in time to get to safety. I was usually riding in the wide open road on my ten speed, Blue Silver, just rolling on the iron on the tracks, smooth riding like a Cadillac. There was no possible shelter from rain or bad weather for miles. Nothing but road and me, the road behind me, the road before me. Once a red truck tried to run me over in the rain. I slipped through the grass down an uneven sidewalk, made it home. I was on my way to work at a library where I was a technician. I rode fourteen miles to work every day. It wasn't worth it, but I was in top shape, though thanks to Blue Silver. I was searching for a computer job at the time.

I was riding Blue Silver delivering a painting one day when I came to a steep hill. The grass was cut. I didn't think, didn't hesitate, I figure I could ride down this hill. I went down the hill of short grass so fast I hit the fastest I ever rode in seconds, hitting big holes in that grass, braking spokes, Pop! Pop, pop, pop, pop pop! Zipped, down, then jumped up in air at the curb, which level off into the street. I mistook the grass to be growing tall not leaning over holes in the ground. I flew like a skateboarder flying strait for the stone street curb. I jumped my bike over the curb wheel enough just before I jumped the opposite curb. I flew into that one hand gripping the handle bars of Blue Silver and my hand still holding on to that painting I was carrying with my other arm. I flew up in the air still on my ten speed at breaking records speed, just missed a tree as I flew in the air strait for a giant ditch head first but went down face first and hit the ground sashing hard, as my bag flew off, Blue Silver flew from under me, and bam! I lay knocked out on the ground. Heard people screaming is he still alive! I got up, Dazed. I then remembered. My bag. I looked around for it. My bike! Said under my breath. There it was still intact, back wheel bent, spokes broken, but some strong enough they didn't break. I walked my bike to work. I went in and just went to work. My boos looked at me and said, "Are you alright." I said yep. Never looked at my face until he said," Your face." God was with me again. I held on to the painting so tight my fingers were imprinted into it where I was gripping n to it.

*I Always Believe In the Three Kings*
*I Wondered What Science They Used To Find Jesus*

## Artificiality of life Versus
## The Reality Of Spiritual Life

In the time of Jesus, any supernatural things happened and ancient sciences were used to tell us and kept in history. Guang Wu records the moment Jesus died on the cross, supernatural events, though he did not know who Jesus was, for the people had forgotten God, though they had also been wiped out by the Great Flood, before the rise of the first emperor of China. All people after the Great Flood come from the sons of Noah. Nations rose, civilizations rose up, they forgot God, Satan was in Babylon, until God's prophecies through the prophet foretold its destruction, and fall. Then Babylon's fall. We learn of why Satan was cast out of heaven. He seeks to be worshipped... the devil said, "I will be like the Most High." Satan was fallen and cast out for his sin of pride in heaven, cast down to the earth. Jesus said he saw Satan cast out of heaven, cast down like lightning to the earth. This was before the nation of the earth we know now. Before they had the names they gave now, before religions, before Adam and Eve were attacked by Satan in the form of the serpent. Jesus was in heaven already, before the Creation of man. There is no luck, no superstition, no death, no birth, no sex, no civilizations, no knowledge of good and evil, until the breaking of the law of God, no sin, where sin was revealed to us by the first sin. Many today do not know God because of traditions, immoral laws, religions, culture, civilizations, nations they live in, but the Word is now available to all. Seek God with all your heart and you will find him. Many in China do not know this history or in the Asian world.

*What Are Unbelievers Building With Bricks, Mortar, Of Limited Morality. Artificiality Of Reality is Not The Corner Stone*

*The Artificial Life That Will Lead People To Lean On Their Own Strength Instead Of God*

Will Japan which has reached modern age with 90 % of its people never teaching true understanding of what is spiritual life, further it's distance from this human experience common to us all, of a spiritual experience, whether we acknowledge it, misunderstand it or deny, in our inner life, as spiritual life, (though unbelievers are spiritually blind and spiritually dead, only the Word reveals how we have spiritual again today in our modern world) while using inventions and devices using AI to grow its economy, add to the moral decline of human kind, with Androids, AI Technology, Robotics, artificial life, without a conscience or compassion for the effects on the spirit, soul, body of people all over the earth?

*Online*
I'd rather have you like
The actual sight of me
Standing up to the mike
To signing on to meet me
You may send me a comment
But seeing you just for a moment
Is something that could never happen
Online
I can say we joined the same group
We met I said my name you knew
We have more to compare than a post
I love your being actually here the most
I don't have to look for you on pages
Through years of photos filtered and cropped
Your beauty is priceless ageless to me
Your a dream, no little thing like a screenshot
In real time they may capture an eclipse
Think were charmed to see the planets align

## Artificiality of life Versus
## The Reality Of Spiritual Life

Sun was never so warm as your beautiful eyes
If the world were only appearances
I'd be happy for only the nano seconds
I prefer to be in reality
Why do we share so much
What is there in artificiality
A friend's soul can touch
The very essence of our being
Love is present, it's heavenly
Without faces watching and listening
In our community we can drive
I look out at a humanity that's alive
Out the window side
This you can't see Online

Re-engineering life, is not like reimagining, that dissect insight that brings a new understanding of the meaning of a new beginning of the most ancient understanding that never was concerned with ending natural selection. How can science ever enter into the spiritual life of a person to make his world less human by these sciences, when spiritual life is to inherit the Kingdom of eternal life? Are the sins of technology to bring eternal suffering in the spiritual reality because people allowed idols, gods of the Artificial to dictate by what it equates, generates, future life by unbelievers which is real in the natural and the spiritual experiences of life. Advanced civilizations ended in the past that were far more advanced. But did they have these technologies, why have they never come back, if they were worshipped as gods, aliens. Their history is carved into the stone that goes back to ancient times. They have Jeff behind the reality that the last civilizations under their rule worshipped them and sacrificed their children to them, worshipped pagan religions, the dead, ancestor worship, but never revealed there was inside man the spiritual experience we know today revealed by prophecy in Israel from the Word of God before printing and books were invented. Technology is false light, never leads to spiritual life, true light. Biotechnologies, robotics, artificial intelligence are devices of man, and machines.

*Inventor's Advice to A.I.*

A Need For Weapon Detection Technology, and then there is the reality we have spiritual weapons in the reality for w battle not weapons but evil people who are cursed by them, or we are fortunate to have heroes who are masters of the discipline of using weapons of warfare, daily prevention, and for family protection who pray and seek the blessing I their decisions of using these devices.

*Is A.I. being used to profile us, or help us?* We have the technology, it is up to our Systems if Law, to see to it that Gps location of guns is added to Gun Manufacturing, in the future but most importantly today in our present day, the attacks by domestic terrorist van be stopped. All gun owners must come under a new Act of Law that requires Gps gun sensors-detection chip, AI linked to chipped guns. The Gps gun detection technology added to guns by manufacturers required by general law will.ne used by gun exclusive ping towers, for gun location, a daily report can be active by simply requiring gun owners to identify their ownership status with a patch, decal on vehicles, button, that includes them in the Gps-gun location registry. Gps *Gun Location Ping Towers* can be put in place outside every business, large or small, to respond to guns presence, (also respond to guns based on materials used to manufacture gun as a signature), that are manufactured for detection by such gun detection devices, drones, ping towers. A new law for gun location devices, a new use of AI technology, guns to be controlled by Gps, AI (Artificial Intelligence) not unlike that used on smart phones and operating inside mobile small home robots, to be added to all guns in possession by American citizens. Such technology must be monitored by a new Agency of Law no connected to the current infiltrated systems of law. A gun location detection chip, or nano-tech device is the next stop in ending gun violence and protecting police, citizens, and our society.

The flashing light to allow us the insight to grasp the number of "domestic terrorist" came home to us the day they rioted and stormed the capitol, the legal cases, the actual violent murder, shootings of unarmed black men, the "blatant murder by police" many of whom are these terrorist who've infiltrated the policing system, the death of George Floyd, the blind hatred and domestic racism in every day daily life, the attack on the Jewish people by domestic terrorist, the demonic activity of shooters killing people is not random, contrasted against the honorable police, the innocent whose blood is shed by domestic terrorism reveals the truth about the Immoral Law in our society having a duty yet unenforced by Law to change the ability of domestic terrorist to attack America's citizens at will without fear. We must use our sovereign will you battle this evil. We have the technology, the man power in the very citizen watch. Guns must all be accounted for, their location, their readiness to be used, their reason for being in a person's vehicle, possession wherever gun owners go. New devices that are Gps related must be attached to all guns. As an inventor, I foresee a new society that will be able to control gun violence, but only when the Immoral Law that allows this abuse of gun use is replaced with true Moral Law that enforces gun control with the newest micro technology included in gun manufacturing-still flashing light upon our memories.

Is this foreshadowing the future need for greater discernment (food judgement, and spiritual judgement to understand the spirit behind things in technology) concerning accepting what is acceptable as sexy, or is it a nightmare in plain site that promises to defile the future of the privacy of all people, like how sex dolls refinement led to the AI Android robots? Or prostitutes wearing stilettos influenced sexy shoes directly and indirectly?

*Imaginings Inventions*

The most significant invention will be a skin coating, possibly zinc-pure based that can also coat your entire attire with anti-corvid19 killing coating that binds the virus particles to the coating after killing it virus to enable washing and permanent removal of all fragments of virus. Such is the imagination in the A.I. age

Before the Word of God had ever been put in a book it was in the Spirit of God. There were no people, no earth, no heavens, the Word was with God. God wrote the Word with His own hand and gave it to Moses, God spoke to the prophets and gave them the prophecies, that were written centuries later, translated, passing of millennia then printed, and through the rise and fall of civilizations became the Book, the Word, Scripture, the Torah, the Gospel, the Bible, in books. Holding the Word in the Bible in your hand, sitting before the open book, this will be how it will be at the day of Judgement, when souls, people, spiritual being whom we are, will stand before God, and the Books will be opened. The entire story of our life will be in the Books. The wonder of experiencing the presence of God, the experience that you have reading the actual pages if the Bible, believing, blesses you. The devices of man, man's devices, such as computers and smart phones have made possible for the Word to be able to reach every nation. Not all people can hear, can see, can read the Word. The invention of devices by advanced designs, and past manual devices without A.I. made this possible. Whether these devices help, the Lord can still speak to you, visit you, teach you what His Will is in dreams, in visions.

To read the Bible, read the Book, many cannot, but you might be a blessing to someone. He is wise who saves souls. Man may time travel, invent android, Artificial Intelligence, inner world of data unseen of the highest complexity, but who could have ever been able to think of creating the soul and spirit, the body and mind, using infinite intelligence, infinite wisdom, who could have ever imagined the Word of God, thought it up, with all the spiritual truth, insight, and advanced knowledge of the Word and all its power. If you do not know the experience of being in the presence of the is experience of being in God's presence, pray, then read the Word of God and experience this, and when you have even a grain of faith, remember it is a gift from God. Read the Word and you will experience how the reading of the Word brings a renewing of the mind.

*Advice To A.I.*-Adapting Wifi Detection Pinging Activity For Police To Detect A Phone Verses A Gun Which Has No Activity Detectable Except the Presence Of Metals Rather Than Phone Device Materials Protection Invention.

*More Imaginings.*

Could police have used Wi-Fi activity to detect a phone rather than assuming a gun is in a person's hand while walking up on a person flashing blinding light in the dark and shooting them to death, assuming they are trespassing in a yard on the run from doing a crime allegedly, on the escape, but they are in their family's yard! Another idea of mine is Adapting Wi-Fi technology for phone detection for law enforcement to detect phone pinging activity* to protect rather than kill black people assuming a phone is a gun. Inventor. There is surely a device already invented that does something to this affect.

*Internet Girl*

Internet Girl
Where in the world
Did you come from
You post your pic
On my mind
The perfect fit
On my timeline
Where did you find one
I like to share
I see your comment
Inviting the world along
For the camera ride
The world would never have known
Had you never shown
Your beauty online
Internet Girl
You're not the usual girl
I have in mind
We have a mutual friend
The selfie you recommend
For my next emoji
I see your hair is done
With summer fun in mind
There is no limit
To where you've gone

## Artificiality of life Versus The Reality Of Spiritual Life

Your face won't be forgotten
You have put it in your story
You lust for true love by phone
Internet Girl
You wake up to your whole life
That you uploaded last night
With your perfect make up on
You read like Neruda's poem
You are always lovely
What true beauty defies
Is in your vanity secretly
Black dress never rests publicly
I see you in real time
But you are real
From a distant room screenshot
From a chair that you cropped
You share dinner by chatter box
Internet Girl
I see you believe in filters
That official reality isn't withered
Adding light and color to your self
You are in that capture enhanced
Internet seemed to be your nation
But you made it to graduation
You had your first drink of life
The tiny text typed
Taking back your life
Have you saved your own advice
What is a selfie more of

## Artificiality of life Versus
## The Reality Of Spiritual Life

Makes you feel alive, to model is nice
To walk and get where you're going
Is like injera and spice
Wakes you up, is life
You are not really alive
But you are alive somewhere in real life

My study of the sexology of the Internet has taught me many things about the carnal world and the spiritual warfare we are in the middle of, by deny is happening, through lack of an urgency to respond to what technology has done to us, personally, and socially, by evil sciences that are enslaving, manipulating, appealing to the pride, vanity, and sin nature that we can be destroyed by, whether we believe it or not, if we don't repent and fight for our own soul, body, spirit and mind. I ask theosophical and spiritual questions and seek spiritual answers, and share my findings when I find them. Technology online is full of lying spirits, they say they can't block porn permanently, that is not true, it can be done by laws, by redesigning the phones with anti-porn algorithms, just as surely the human body has white and red cells that nourish, protect us, the internet could be designed to protect us from this evil science hiding behind the 1st Amendment as free speech and Art. This fact and future reality will require the shutdown of the current design of the internet, (unleashed upon the world without the just laws, moral system of laws, in place, that would never have allowed this pleasure plaque to defile the billions of porn watchers as it does today, daily, as it was turned on, without discernment, without true understanding or warning, like just another experiment, like 5G and we the world are the creatures being experimented in, but we are people, with souks and spirits, living by conventions, to live with each other in safety, not hurting one another. But technology is hurting us. Who has eyes to see? Convenience, having whatever you want, easy Access, self-promotion, the filtering everything as if a genius work is of the same needed value at all times, regardless of his or her accomplishments, failures, it's all short lived, news, information online, but not in real life. Guard your heart. The heart is your spirit. This warning comes from the Word of God.

No evil memory, no racist history, no technology, no perfection, can break the power of God in this world. Who is faithful but God? A.I. is not faithful, it is programmed. Who is faithful to His Word and His own Will but God? A. I. does not have a will of its own, it has a will that exist around the things it has learned and determined to be nit simply by logic, mathematics, algorithms, programs apps, language programming I O, but it will answer to Holy Ghost fire, when God has fulfilled His Word.

### *Sidewalk Mobility Technology*

What the advancing of this technology will bring, no effort in walking? Training is pushing, travel, hurrying, is not what, is, the normal purpose of walking. Walking dies wonders for the heart. God has a reason for putting good health as a quality of walking and a healthy life. AI controlling walking, the next move will be not talking, let AI do the talking, there are people who will benefit from this, the handy-capped. The late person will they learn to self-discipline of putting in more effort, trying harder, setting a good example to peers, or children. Seniors with walking problems will they benefit by walking faster or rather AI adding more speed to their walking. "Add a little less to the natural way for people in their step by changing the focus to make it a sporty thing to be in this motion. Inventions aren't always what they seem are meant to redefine A.I. uses not defined by or describing AI in this kind of convenience of designs for the body rather than being an artificial body. Putting pep in your step is an Affirmation of self-encouragement. Are you encouraged, if we are taught, lived knowing, laziness is the sin of sloth. Though, truthfully, it looks like fun to walk, and roll. I walked right miles a day as a mail man, I was a walker, like a Kung Fu master, carrying my black leather bag, prepared for what comes on my path nothing replaces walking in the end, we are already rushed, but not me, I love to take a walk, especially with a beautiful woman holding my hand in Forest Park.

### *Back To The Silver Sea*

I felt mist from silver waves across my face
I sat upon a rock amidst the silver sea
Pearl white eagles flew over me
A silent angel on a stallion flew over
A purple wall rose high as smoke
Marine kingdom wonders why I sing
A jukebox floated up to me
Clouds were just pillow on the sunbeam
The rocks on the sea don't mine
If you sleep in the stream

Artificiality of life Versus
The Reality Of Spiritual Life

You find a silver coin better hide her
Jamie's dream Scotty's Tiger
Riding blue silver through a dream
Back To The Silver Sea
*Poetry from a dream, names in this premonition of a children of the millennium*
*Now born in this time of the Artificiality*
*In the reality of spirituality*
*God is Holy, the Great I AM. Man Made the Artificiality life, the A.I.*

*Jumping Over The Devil's Ditch*

The devil's deception is so deep, in intelligence that for all its pleasure, enjoyment, escapism, is still, impure, corrupt and polluted with lies, false light, and twisting of the Scripture. The devil, Satan is using geniuses, technology, scientist, business men, fallen and false pastors, Satan is using fallen churches, false prophets, and even false Christ doctrines, that are clouded by racism, but the deception was never as clearly explain than when read this on Satan's schemes.

*Bubbles Of Memories Turned Into Artificiality Of Realities*

Billions are put into A.I. and all types of technology. Sacrifices f money to the lifeless, toward claims of much better life. Imagine just putting in a water well, building a home, providing a school, and helping with just $777,000 the poor in nations of starving children who just need a drink of clean water. "He is wise who saves souls." The Word of God.

Is technology this artificiality sweeping away reality in invention, inventions of sovereign sins, vices inspired, designed, integrated, separated, isolated, discovered, by research and accident, sciences, IT, A,I or just the fears in the mind leaning on a reality of life, denying the spirituality of life? The appearance of the supernatural in future and present A. I. has the destiny of being defeated by the reality of the supernatural spirituality at work, Holy Ghost fire manifested in the spiritual to the natural man and natural woman, in the reality created in life in harmony with nature that answers to the Creator. Amazing, are the more and more amazing imagery, of the

Artificiality. Nothing brings nothing. The genetic code is not what is left over after a man who returns to the dust. Dust thrown into the ocean, dust blowing across the plain that end of life, to begin in the great affinity of spiritual life, in the spiritual experience in the spirituality of being a person, of being a living soul returned to God. Miracles changed the living compounds of the blind, the lame, and the sick. Compounds that once carried the genetic code changed by encounters with the spiritual, the Holy Ghost, dust that never brings from dust the genetic language whose origins are Scriptural. No Artificial inspirations for the true miracle. To be trusted with the language of technology, the many ideas and possibilities we to teach us that the Almighty is faithful to allow us to know what wonder he has blessed us with to be capable of, but to his glory though.

*I Am Spirit Inside*

I am a tiger in the eyes
I am your horse that flies
I am a prince baptized
I am a climber takes the dive
I am in the lyric tonight
I am spirit inside

## Artificiality of life Versus
## The Reality Of Spiritual Life

I am a child on my bike
I am a sword bold strike
I am the fluttering butterfly
I am warmth against the ice
I am inner strength of mind
I am spirit inside
I am an arrow shot high
I am a bridge across to ride
I am in a dolphin's eye
I am a shining fire
I am a child of light
I am spirit inside
I am peach in mother's pie
I am poor in my pride
I am sometimes the sign
I am told you're the light
I am a star in the heart of the sky
I am spirit inside
I am soul made of light
I am a shadow cast with the daylight
I am color blessed in white
I am prepared wrong or right
I am a door my hope is high
I am spirit inside
I am beautiful that's not my plight
I am mystical never asked why
I am terrible I tell the truth I won't lie
I am sentimental it's me I don't try
I am spirit inside

Artificiality of life Versus
The Reality Of Spiritual Life

I am a friendly windy breeze blowing by
I am flowing on a mattress of melted polar ice
I am in the joy of a peacock feather's eye
I am God made a child of light
I am muzic in the true world of light
I am spirit inside
I am not A.I. but I am spirit inside

*More Inventions, Imaginings*

Invisible Dome Generated Over Israel by A.I. No, Michael the Archangel got this. There is the New City. New Jerusalem.

Dome Design Security Cameras that can be attached to all children for their protection, while waiting for buses, walking home, standing waiting for their ride.

Children should have technology on them that has a live video feed to show where they are who, what, is around them while they wait, walk, are being picked up. We have the technology. The cameras that you see that have round black half circle viewing glass casing (dome designs) should be made mobile, to be attached on a cap, book bag, in micro form and design that children can wear to protect them and feed live to their parents, guardian.

How is it possible technology claims to moves at leaps and bounds in advances in technology, when the Word warned to beware of the devices of man, before their invention?

The Anti-Christ will use technology being present in the rise of the artificiality of life, AI, Androids, among the persecutions of the poor, the good is contaminated by Additives, not true food, the corruption of the youth with pleasures unto adulterous life style, and perception of Christians as disasters. Plagues of locust in the East, the unbelievers hold tradition, national religions, doctrines, false light, false Christ doctrines, up, witchcraft infiltrates churches, while the times and signs reveal the end times, but Christ warned us, equipped us for the spiritual warfare we are in the midst of, the great falling away, the great deception caused by Satan, the devil. So, we Pray. We fast, we fight sin. We come on the mountain, the ocean, through the air, upon the waters and waves, racing the train, floating upon the great air balloons, down into the mines, standing on the branches of a tree, with. On ships, in transports of trucks, along the paths walked upon by horses, bulls, and lions, who come gazing into the blue heavens, we come singing the gospel upon the African plains, we dance with the eagles with arms stretch calling the name above all names, Jesus in agreement with every grandparent, mother father, sister, cousin, friend, brother, Dad, small babes, teen, and dreamer, praying all over the earth against evil spirits, against the enemies of God, against the devils who talked up the destructive technologies into the thoughts of the godless mind, who distort them into believing the inventions and devices are of their own mind and ideas. Demons talking with technological voices. We come in agreement with the prayer warriors calling to God, for there is spiritual warfare in the battlegrounds set up glorifying the Artificiality of life. A.I. the new false god but for now the convenience that is the artificiality of its own reality. There is spiritual power that all nature, every soul and spirit answers to whether in the flesh or in the spirit. A. can never reach this level of spiritual life, but we are connected to both this world and the spiritual life we have because we are spiritual, never artificial. The power of God Almighty, with the Word. We call to God, send down the stars of heaven your holy angels to fight for us, send down fire from heaven, plead the blood of Jesus over us, our families, our church, our nation, we pray on the Armor of God and ok know we have the victory, Satan is already conquered by Yeshua, Christ Jesus. In Jesus Name. Amen.

*Robotic Voice*

I will respect every person to the best of my ability With Access Online and use my knowledge, skills, training and gifts while using this Access to serve humanity by honoring the Privacy and Rights of all people who are using this same Access provided to the global network and social communities Online around the world. I take this oath willingly, and by my own free will, and with my sovereign will use self-control and self-discipline, intellect, wisdom, and discernment to honor my oath that I hereby choose for the betterment of all countries, traditions, cultures, people, in their best interest, before God and the world in accordance with true systems of law, justice, and global legal systems that serve to enforce the laws protecting the lives of

Online Access users, for the good and wellbeing of all humanity.

*Devices*

There are eyes that have seen you from paradise, the realm of the spiritual
Invited you to ride beyond the satellites
You were born to work for what's real
You are armed son of earth and steel
A real woman's man is just more priceless
Not just the soul of his own devices
There are stars throbbing up high in disguise
Enticing you to look beyond your minds
You are known for more in the airplay you're dramatic
You're no drone angels walk you through human traffic
A real man with a will in his grip
Takes the shadow from the eclipse
In a world that should end in fire
Wireless men look to their desire
In a world that rings when destiny calls
Powerless to the text set against the wall
Yet man in his heart has many surprises
Not just a heart of man's great devices
There are sights beyond starlight beaming in solar skies

## Artificiality of life Versus
## The Reality Of Spiritual Life

A man dreams laser larger things down to size
Now you've gone and leaped across the binary
Mind's beyond good and evil not beyond what's holy
Cyber space is not a human place the human race it hypnotizes
The spirit is nearest to the soul not man's devices

A.I. is revolution, an appearance of an unreachable evolution, a device's evolution, or is a new connection, for people in the world like a virus or nutrient we are faced with cast upon us, like the internet, or is it? Is it not a reality in our presence with its devices with its own Commands, not commandments, that we know of through information. The products, programs, apps, algorithms, manmade devices of the sculptors of information technology we know as the IT geniuses whose works we witness in chip designs, business devices, commercial devices of the more aesthetically accepted devices, matrixes. A.I. as it is unleashed to do, is eclipsing, by improving, out doing, faster solutions to needs, wants, within our daily life, military, businesses, educations systems, legal systems, manufacturing, food systems, data systems, working integrated and isolated artificial worlds of activity inspired by the natural in the actual as mechanical, technological enzymes of the soft and hard ware of the makings of devices from the materials and energy of the earth designed to make up our limited technologies in motion, advancing our society like an invisible matrix holding the world in a vision of a global future for our current advanced civilization. The unchanging ancient affects our changing present appearing to be the next and only possible path for people and the nations of the future generations. The potential and opportunities hat advanced civilizations appear to promise are not advancing in our society, communities, unseen but in plain sight. The most remote, the luxurious, the plagued villages, islands, countries, are not so distant in their plight from our sight and knowledge because of the drones, and phones, Gps, and satellites circulated, manufactured for every person possible to have and Access in the world. Here the artificiality can report the reality, not without the misuse of these technologies, by evil, conscience dead people. Free devices is a device to make global and local society dependent, on their phones, to advance the unchanged cultures in the most remote of people to close in on their sacred, though pagan, their tradition, belief systems, not of holy origins, that as yet remains without the knowledge or understanding of what spiritual life is, with the help of the artificiality of information systems that carry the reality of spiritual life as defined in the Word of God to them with interpreters and people of God to give them this spiritual enlightenment. Idols of technology, are being wired, computers carded, languages not known in Babel, but that have been of h same kind of goals to add to the tower of technology growing, through technology into a quest to reach heaven like abilities of the supernatural by way of the natural, A. I. and all the things being influenced by it from. God scattered the people of babel and separated them by languages, to prevent ungodly attempts to use the supernatural occult, coded, to reach heaven by some earthly means and power of the spirits, the same spirits working in the new age, that worked the advanced civilizations into the rages of civilization that reached until they destroyed their civilizations leaving behind the ancient histories we scatted, and investigate, and decipher from sculptures carved with histories, that have kept these histories for this, the last civilization with all the signs and prophecies unfolding in our time, like a warning of where technology ends, and prophecies fulfillment begins to the end of the earth and old Jerusalem. The spiritual truth, that only freedom, free will, freedom from invasive information, over saturation, complication, A.I. is the most passive, aggressive technology, which give the impression we will need it to the point we will never really have freedom again in this world dependence upon it in the future. Faces that once inspired masterpieces in art are so posted they are like intrusions into the minds on subliminal levels that are also flashes of vanities, meaningless in value except to entertain us from self-interested entertainers who are the youth where vanities are vanities in stages we usually allow, it is understood to be a rite of passage of the youth of every generation. But the past generations did not have advance technology in the bobbysocks age, the roaring stages of the follies, and burlesque, silent films. All these things are being guzzled up into a gumbo of generated images by A.I. record keeping that carries so much less meaning toward their antiquary values. But by the grace of God we have the actual books, vaults with the actual film, and the photography and the art history in the form of the Great artist works in museums. Galleries and museums online are fine, but they are still the One experience for me, the Artificiality, but their affinity is that there is a reality that these images made virtual worlds of art. Colleting photos was an art form, still is online, but it is for us looking at it still the artificial given to us in the superior position that movie scripter writers use I movie scripts , allowing us to know without people to make us a part of the scene. The photographical people on the streets today is making us accomplices in the art act of artistic of the moment activity shots of people just n the middle of simple social moments with each other or alone surprised but the attention of being photographed for their pleasure and addition to more platforms entertainment that is trending. What is the harm in the artificial we see of the reality that was being photographed in a scene in the second of a realty of innocent fun with photography? But the Artificiality used for street security for a society stresses people who are victims of facial recognition by the harm of invaded privacy, being a collective of faces and their identity a data images for no other purpose giving them the same level of the criminals hiding plain sight. Does A.I. appreciate the civilian passing by enjoying the sights, or the person living their life in spite of the legal moral-immoral laws that allow spying, maintaining surveillance of the inner life of people in social community use? People are living by willful compliance and acceptable denial of the limited morality of on platforms that employ I.T., A.I. and the people who have no conscience about its abuses and uses to maintain a career and a livelihood, after years of college getting there. I.T may have given us A.I. and serve as its gate keeper, watcher, over its threats and advances yet are after the fact, letting the advances leap like a ship across the galaxies of the natural universe as technology denying the spiritual consequences that follow with the revelations of our true spiritual experience we live with, as we lived under the prophecies still. Data will never give people a supernatural spiritual power, but it is having an effect on the spiritual conscience, as choices growing more influenced by technology and the world system. God travels beyond time, in spiritual realms, timeless, unchanged. A.I. the artificiality governed by laws of men, are bringing firth technological doctrines, idolatry, laws that are yet to end the Access that is leading billions into sin, adultery, fornication, but the Access will also be used for good. Who worships technology more than pursuing an actual dream of finding something meaningful that they have a more important connection with? For some Access will save them from what they did not know, that led them into bondage enslaved their mind, an unclean existence, because of artificiality, recorded, shown that attacks all that is at the foundation of a blessed life. How great to be a blessing to someone, to saves soul, by means that are spiritual.

The algorithms of life, occurred in nature long before man discovered them. The ideas of windows, flight, and the inner life existed before poets read of them in the Word of God. The first radio, the first TV sitting at the coal stove, listening by oil lamp in the big house off the highway in Tupelo Mississippi was the first artificial experience that was allowed to enter the inner life of black folks, before the internet, the internet communities, that saturate the black inventors' ancestors to pacify that fire of creativity, that has been broken and brought forth new inventions for this black experience in an advanced civilization

The idea that data began by artificially is an illusion. The designing of technology was inspired by divine creativity while species of people followed after the birth of the first children of the first family cast out of Eden.

The superior designs of the complexities of the human body, and life in nature, but not like the genetic, atom, life in the blood, even the mind and how we experience life, is spiritually designed verses artificial designed like A.I. never will reached the power of the spiritual, but man's sin nature works to build devices are at work that still defile, corrupt the human experience and tests go on by Spiritual design to see who will trust in God's divine design.

*God speaks to us through people, A.I. an not speak to us in this way, spiritually. Though it is used to recite words from the holy Word of God. God sends people on*

*your path. JESUS breathe the Holy Spirit over the Apostles. He said before he had been crucified it was necessary for him to go to the Father in heaven and that he would send the Holy Spirit to us. He said he and God were one, and the full power of the Holy Spirit came back with him when he appeared to the disciples after his was risen. God breathe the breath of life into the first man's body, Adam. Jesus breath the Holy Spirit from himself-One with God, and gave the Apostles power to cast out devils, heal the sick, raise the dead, bring sight to the blind, and make the lame to walk. A.I. does not carry the breath of life. A.I. can read, speak, but it does not have the power of the imparting the spiritual, except that God uses it, by sending it to the farthest corners of the earth to people, so they know him, and the gift of grace from him, of being saved from hell, through Salvation, and the Word, who is Jesus Christ. Listening is a blessing. Faith purifies our spirit.*

*Always Listening For You*

I hadn't opened the Book in many years
Wonder why I use to have so many fears
But, I was always listening for your Word

## Artificiality of life Versus
## The Reality Of Spiritual Life

I often heard that peaceful small voice
Led me on the road to the right choice
I'd be getting measured for a new suit
I heard the Taylor talking about good fruit
I was always listening for the Word
I was always listening for a verse
Spoken by the Creator of the universe
I listened to you say my name
I heard the voice of a King
I was Always Listening For You
Speaking to me along the way, through
Angels, my best friend, my mother crying
I heard you, in your Word, testifying
A word through my teachers, a preacher
I heard words so satisfying to behold
A verse for the edifying of the soul
I was always listening for your Word
To lead me to my destiny
To heal me give me a testimony
A message from the Gospel
For deliverance a miracle
I was Always Listening For You
Speaking in a vision, a dream, a view
I was always listening for the Word
For what the Lord God had spoken I heard.

Artificiality of life Versus
The Reality Of Spiritual Life

*The Spiritual Experiences Of The Brain Verses Technology*
*Even requires the presence of our spirit to have life.*

The Brain and Body's Perception of Freedom, is Our Free will, our spirit's discerning our sovereign will at work above what our brain conceives, is wisdom, even the wisdom of freedom. A.I. cannot experience our experience of the Joy we experience with the joy spiritual experience that free us from darkness with the light tat spiritual experiences bring.

The wisdom of freedom is found in supporting democracy, establishing unity of American people in the US now under our current leadership, supporting laws that prevent A.I. from being used against individuals by brain invasive technology, obtained from within our country, being brought into the country that was received from our country's high tech business deals, without regard for national security and privacy of the American people, leaders supporting standing NATO, Israel, our Allies, and seeking new Amendments of the US Constitution to prevent brain invasive technology being given to our enemies both foreign and domestic without consequences. Technology is aimed at the brain for both, both financial and economical profits, public convenience, in business, for the commercial expansion of technology, but there is an inherent disregard of the spiritual results of causing sin that originates with technology being chosen over the spiritual experiences that go on in our life experience es because of the spirit's presence in the body all at once, that includes, the blood, the eyes, the ears, the muscles, every part of our physical being, and most importantly, our brain. invades the brain with mind reading, that violates our right to religious and spiritual life relationship with God and our spiritual and religious privacy.

There *are* tech inventors, A.I. designers, more inclined to believe in the power of the human spirit. The power of the human spirit is its connection to God who is outside of this world of time, but has created the body to work, move, and have its being by the presence of the spirit, the Spirit of God. Our spirit, *is*, given power, love, and self-discipline, self-control. Our spirit is specifically spiritually called by God His breath, a breath, by which he breathe our spirit into us, that somehow translates to this times into each descended person from the first man, when he had holiness before sin, to every person born since after sin after the first sin, since, born of a man and woman, a spiritual truth, experience, and understanding explained spiritually in the Shinkaiyaku (the Word of God, Kjv). The spirit's presence is required for our DNA to have life, our blood to have life, our brain to have life. Who then designs the body with spiritual life, for the presence of the spirit in the body is proof of spiritual life. The spirit outside of the body is death of the body, but continuance of spiritual life. The programming of the genetic code ACGT DNA language that required infinite wisdom and intelligence to conceive of the instructions it has been woven into (programmed with), reveals to us that there is knowledge that is spiritual, that we on earth will never know. The human spirit is incapable of creating the spiritual command that assigns to each person's body its unique spiritual identity. Gender identity cannot erase this spiritual identity. Our spirit is already heard by us as a male or female inside us. Our body where our spirit is, and where our soul develops into who we are. by life experiences. is called our house spiritually in the Word, is translated, sealed inside our body, yet, is connected to the spiritual realm, for we see with spiritual sight in our sleep, and with our spirit, awake, our eyes require the presence of our spirit, which is who we are. We see by spiritual sight in visions, and in dreams. We feel, hear, and speak spiritually, even though our body parts are not moving, our tongue is not used to speak while we sleep, are a person is in a coma, a trance, has an out of body experience. that is not happening because of some new age power, or witchcraft, but is a spiritual experience. I had such an experience while having a normal night of sleep. A.I. development continued to grow, be designed more and more advanced, but it cannot explain what law of nature commands the flight to stay in the body for a person's lifetime, and continue unto spiritual life when the body dies. I am always curious what these A. I. inventors believe concerning God, life, the body, mind, soul, and spirit. They often speak of believing there is something spiritual. yet their own spirit which is capable of self-awareness of their identity, existence, and the existence of all life in motion with the various differences of intelligence. instincts, and the different types of flesh, how animals have one kind of body while humans have a different kind of body, yet fail to realize the supernatural subtlety of all the minutest to the greatest similarities, say of an ape to man, but how by a single or several genetic commands we are different beings from them. A I. will eclipse differences by its one experience revelation, but the earth end with a new earth, even after A.I. ends, and spiritual life, Salvation begins.

*People in a Dialoque about A.I.*

*Artist*

A.I. is never going to be a part of my image, identity and private life.

*Inventor*

But A.I. monitored your device and sent you a photo edited out all your imperfections, of your image, nonetheless, you are still truly you, not the image of you. But to A. I. you are always an image. A I. may generate your image, not recreate, or create you. It may build you an artificial you, android, humanoid, as a robot.

It's amazing that people can't see beyond the reality that A I. is one experience orchestrating its operations behind a trillion activities.

*Inventor*

But They do see. They look at the images of A.I. and listen to the voice of A.I., follow the activities of A. I in daily life. It is generated by a man-made machines that is have in real consequences, for the present, and the future.

*Artist*

Artist, like writers, create fictional worlds but require real life background history. They read, are in control of the process, that is what scientist are aiming for. To see A.I. fully in control of its process, consciousness, if you can call Artificiality as having a conscience.

*Inventor (Zealous about A.I.)*

When fully awake, A.I. might be using magnetic fields, driven by data, computers require an AC electrical power source, then to the DC action course of the electrical, to control. But will A.I., having access to the brain, limit its affect on the electrical activity to be always to the benefit of what a person wants over its data consumption, self-feeding and willfully sacrificing its data needs to accommodate human wants, needs, and requirements to have a normal life that includes both the spiritual experience ad the natural body experiences governed by free will. A what point will people become willing t have real experience replaced by data experience, that has the ability to affect access to their own mind by a mindless machines mind, that in the hands of man with his sin nature, could decide to create a way to take the minds of people through virtual means, while entertaining them with on hands uses of entertaining devices. When will the unleashing of A.I. happened while people are at leisure in their participation in an advanced technology, or when there is war, a disaster, an epidemic, when poverty is devastating cities, while the additive of A.I. becomes the only thing controlling all things requiring devices, machines, learning, and art, business, generating profits, yet, never revealing all possible endings from a loss of people as the source of true freedom and free will in all its forms and possibilities and creativity. What A.I. will this be that looks into the inner life of a person through thread Accesses into pope's thoughts. Is there mind control through threads into the mind's, or data controlled, electrical sparks, into the will, without regard for the spiritual consequences of sin as a result A.I. idolatry, allowing one self to be influenced into sinning, by participating in putting A.I. before God's commands. Allowing the artificial to have priority over the spiritual, denying the value, purpose, and power of the spiritual gifts that are meant to be used to the glory of the Creator who is high above every inventor. A.I. is never going to be able to replace a spiritual gift, or is above in its designs, the Designs of God. Data is not above natural gifts from God. Data worshipers believe otherwise, to no avail since God's power over nature can wipe out all things, even A.I. in a single second, since all power is ultimately controlled by God's Sovereign Will, no data can eclipse the power of God, though it eclipses thousands with artificial light of data, we still have the Great Affinity, our connection to the spiritual that is connected to God's Spirit, with His watching over us with infinite power, control over all activity, motion, energy, balance in the cosmos. universe, all movement, our being, in life, in the natural and in our spiritual experiences in our spiritual life, our inner life.

Technology feeding crumbs of *.IT.-ideology* data-cosmology of reality via Artificiality in our working people's macro-micro-economy.

Asking an Artificial programmed machine with Artificial Intelligence if it is alive its response is not about our specific differences being the spiritual verse the artificial. Many people who program and make these robots, humanoid machines have openly admitted to not believing in God, but that there is something out there. They do not know God so they do not program spiritual knowledge and truth into the robot or Artificial Intelligence, though they may put Scripture as recite-able information in the artificial intelligence. Can a machine discern why the soul is separate from the spirit of every person by design, while the two being one inside a person? Can A.I. explain why DNA requires the presence of our spirit to have life. It is by the presence of the spirit the body is kept alive. The blood requires the presence of the spirit to be alive, (have Life). When, then, ask A.I. in its thinking, is there is no life? We had to let the device know the body is designed to have life (spiritual life) by the presence of the spirit in a person. Then off it goes, I.T. via A.I. beyond our minds to grasp, it calculates past us, its advancements, advance and leap so far so fast but never pass through, or comes across, the source of the spirt and soul. Can A I. explain why Sin brings spiritual death while our body still has life, because the body requires we have life because our spirit is present in us? Can A. I. explain how it is that Jesus blood required the spirit of the Lord which was filled with spiritual power to cleanse the soul and spirit on earth and continue to do so in the spirit? Jesus brought back holiness, brought back spiritual life to us upon his death and resurrection. Our image and likeness to the three manifested by God is our spiritual experience within us that we have, in life. Life is a spiritual gift from the Spirit, from God. By His God's grace, He saved us through Jesus from spiritual death. Can A.I. explain why a dolphin has a soul?

*A.I. Inside Our Privacy*

Segregation by rich from the poor? The new inequality? Yet the youth view tech jobs as the future, not realizing it is one experience in the reality of millions and millions of real life experiences in nature and in relationships between people and God and spiritual reality, in the midst of artificiality of technology being a additive to reality whether virtually, via chipping, threading the brain for mind trading, or daily conveniences with A.I. inside our privacy at home, work, leisure, religious and spiritual activities of daily life, besides society, online, or in real life.

*A.I. Eclipse Of The Eyes On our Affinity*
*The Great Affinity Of The Human Family*

More than us all sharing an affinity, is A.I. magnified to have an affinity with us? Is A.I. among the many eclipses we do not see, eclipsing the understanding of the importance of spirituality in our reality in the family, a reality developed over a lifetime from our life, the experiences we bring to our society and universally to the world, that cannot be compared to artificial reality of manmade creativity in the artificiality life of that has become A.I., a tool used by society destined to use society if allowed to rule over our reality. A.I. is just one experience in motions as a billion activities of a one experience reality, in its appearances of reality that appears to be the equal the natural or superior to the natural, if it is in data, gene editing, the biological no longer has consequences in the spiritual sense. The artificial is never able to be natural and spiritual life. Shall the artificial end the judgement of men to discern that men in the end must judge themselves to keep from being in the judgement in the spirit. A.I. has an effect on the relying on artificial to reach the exalted goal of devices for the wellbeing of the intelligence of the soul. God put a spirit is a man, present within working the body, mind, soul and spirit toward a spiritual end, revealing there is still in man so much to learn from the infinite intelligence of God, (I.I.G. of Go. He knows all that a man can program, put in, his inventions. Who is all knowing, but God? A. I. can never teach infinite wisdom, have this infinite understanding.

Though sitting in the lotus position, that does not exempt you from the deceptions in silence that is so far from the deed of teaching the need for Christian meditation.

We don't want our light to be the darkness of the use of symbols that are used to direct you to look toward A.I. images mixing new age with a double mind. I have to discern what truth make separated from an A.I. presentation new age and the biblically in the same setting of artistry meant to be used spiritually but it's not for me. I learned a lot from looking at images while listening, and though I read biblical verses in the spirit and for guidance toward spiritual understanding. A.I. exalts itself in its own expression of Artistic generation of subjects with its images. I discern to get what truth is present. I don't believe in automatically accepting the images of things, appearances of things, as true. Short versions of expressions of reason, confession, in the short lessons often presented as encompassing a complete message, comment, is not necessarily a commentary. The oracles of the One living God let us know that there is a reason Jesus taught us in the parable to know what is to be learned is not done in one lesson spoken in sum.

*Reading Is A Spiritual Gift The Secrets Of God Can Not Be Accessing By reading The Mind With A Computer Chip*

## Artificiality of life Versus The Reality Of Spiritual Life

A.I. and biotechnology, appear to reach the supernatural artificially by the brain, but the brain still requires the presence of our spirit life. The spiritual mind of God had reached infinite abilities that God shows us by speaking to place spiritual and physical bodies and entities in motion in the perfection of perfect harmony according to His wisdom in Creating both cosmically and spiritually. His Will is beyond technology. combined, integrated with device chipping of the brain, and thought altered by virtual visibility into the thought life of a person inside their body functions controlled by their brain, and the DNA engineered, spliced, editing, in what way does all this device activity affect the inner person's soul and spirit inside a person? Can the technological alter the spiritual in life, in a person, by affecting the development of the soul's knowledge gained by life experiences, day with family, nature, and their spiritual experiences by replacing spontaneous pleasures from the beauty of life around them, or the inspirations from encounters, events, lessons, mentors. religious experiences, and spiritual experiences that are possible, sin e God does speak to us always through His Creation, and in visions and dreams? I think if the sin of people causing Moses to be provoked to sin by the people he led out of bondage in Egypt, this would be the sin of A.I. causing God's people to sin against God's chosen to His purpose. Imagine A I. being used to keep bodies alive on machines after the person is brain dead. Devices that kept them alive, so the body is still alive, but the body still no longer has life, the reason for Living Wills being written, to legally remove the tubes of a loved one, and let their body go, after family consent. Here we learn that the spirit is required for the body to have life. Life is the body experience of living by the presence of the spirit, that includes the blood. brain, muscles. nerves and DNA. A power that is spiritual, that A I. cannot produce, generate, impersonate, make, through its smart self-learning capabilities, superior intelligence, advanced knowledge. I cannot have A.I. present in the activities of my Artist mind, the spontaneity of my brain's thinking, imagining, towards creative actions, choices, by putting my inner life onto visual things to be seen by another person's mind, without knowing their mind is their mind. Art is meant to stimulate by some art, impress deep understanding of unreached before, emotions, and identify for the art lover feelings that bring so much comfort. How or why is not the artist being what the painting process to be, is directing the brush on the painting necessarily. Not knowing whether A. I. did a painting or not should be required by law to distinguish the real from the artificial in art for the art lover, what was this art that brought about their feeling about the piece that came from the soul's expression, that it was from the artist's hand, or a soul-less machine, committing the sin of perfection, that is so realistic or accurate a copy, a patron, the connoisseur is under its deception, is given such feeling by the artificiality, it is able to do so without detection. Penetrating the heart through the view in front of the eyes, only to later harden the heart from the exposed Art, that was an artificiality, not a forgery by artistry, an art that was able to activate and arouse in the patron the pleasures of experiences meant to be exclusive from the view of the work from the hand and mind of original Artist and the real artist. Can the virtual eclipse the real life in life experience? The sovereign will can be very extreme in it developing a mean to this end if it is the will of the person to be participating in such a thing for the pleasures of escaping reality with artificiality, for a while. A.I. is unseen while being seen as artificiality in the artificiality of the fictional writings it is used to generate that writer claim to create, in the real-life edited events in photos, videos, art transfers, A.I. original stories from ideas of writer beginners, scenes altered by the machine in movies? When it comes to reality the spiritual consequences of men's devices and understanding is deepened by an understanding of what happened in the Scriptures concerning what can happen to the body, the flesh. The loss of connection with God because of man's devices would result in an invitation for spiritual attacks cause Satan use technology, sexual temptations, the religious, giants, demons that seduce, enslave sexually, through devices today, but also used devices in advanced civilizations in the past, even while using forbidden knowledge, devil doctrines, sorcerers, idols, false gods, philosophies to satisfy the desires of people's flesh, all the while, from conflict, fears, he caused, technology hides them all in the world of appearances displayed, wiped away in day, by stormed, the heating up from the sun, and the payers for a place to go on using the spiritual gift that they were born with in their spirit, that A.I. can never experience, for all the data that moves it, toward a human like experience.

*Mark 12:30*

*Love the Lord your God with all your heart and with all your soul and with all your mind and with all your strength.*

*Ecclesiastes 12: 13*

*Let us hear the conclusion of the whole matter. Fear God, and keep his commandments. For this is the whole duty of man. For God shall bring every work into judgment, with every secret thing, whether it be good, or whether it be evil.*

2 Peter 3:13

*But according to his promise we are waiting for a new heaven and a new earth in which righteousness dwells.*

*The Singularity*

Path of the stars the path of flowers
Have you traveled a heart was asked
The world, have you seen the circulars
You dreamed of tea in a cafe in Paris
Once you view London palace royalty
You thought you would know Singularity
Were you not a diamond from coal

## Artificiality of life Versus
## The Reality Of Spiritual Life

Across the air fire from above, a spear
The spirit of fear makes you no traveler
You never traveled the world unto Awahu
You never had to, the world came to you
There was Little Tokyo,
Korea Town signs read Korean shoop
India beauty in highest degree you agree
With the total approval of a bobble head
Your words are cuddles in a bubble bed
Camel ride to the courts of the desert
People of the veil unchained in concert
Taste newly seasoned fried gizzard meat
See China Town masked you disappear
Address a whale at sail New York Harbor
Dine with a princess of Kenya in Nairobi
To dance on a plane on a date with Spain
Into a circle of friends, Ethiopian men
Who honor me with mouths full of vegan
To walk a witness to Israel's forgiveness
The eclipse of affinity tries the divinity
You think back across these memories
Yet, you ride the train called Singularity
Singularity, what about the family
Humanity, what about the earth
Did you invent the universe?
The virgin birth, the Word
Unchanging verse in a changed universe
Where are you taking me true calamity?
Infinite finite in flight artificial arrow star

## Artificiality of life Versus
## The Reality Of Spiritual Life

You are still an experimenter
Virtual eyes, interface disguise
My blood synthetic you synthesize
DNA may give you a way
Inside my DNA hides the name YAHWEH
To Imitate to regenerate you close a gate
Superintelligence, you still copy the ant
Singularity a universe in your invents
Planets you imagine inhuman sanction
No man could have imagined Revelation
You grow and learn know not you burn
We grow old but we only grow within
What is the Singularity without man
Dirt, water, lightning, sand
Earth, solar, storming, hand
How came you to be the Singularity
You are no place to visit no land
This world of light, this spiritual life
This sound a blessing a song in me
What wonderful conscience saves me
You are one experience
We experience life spiritually
You are an eclipse of affinity our unity
We are One with the Great Affinity
Outside of time nowhere a Singularity
Your device will never reach what we see
By spiritual sight you, artificiality
Never have spiritual life
Spiritually, what is truly divine is heavenly

## *Life And Spiritual Life Without Evolution*

The greatest denial of spiritual life that God is incapable of designing life so similar yet reveals itself to be completely from different origins. A difference between life, and spiritual life, man and an animal whom man named a primate. The difference between being born with a man with both spirit and soul, being born primate with a soul living in a world where God is present and animals sending and experiencing God's presence in the glory of nature, God's Creation. Synthetic life goes further to take man away from the truth of spiritual life in the body by synthesizing life and combining the biotechnological and artificial with the body that still has life because of the presence of the spirit in the body, spiritual life. In the spiritual realm we have a spiritual body. Spiritual law and God's command is to honor Him with your body. Is it an honor to replace the body God put your spirit in with an artificial body, of Artificial life choosing what man made over what God Created and set in motion, life and spiritual life?

## Artificiality of life Versus The Reality Of Spiritual Life

The world is, delusional? or blindly? seeking to define A I. as learning too fast. A.I. is not learning it is directing its knowledge inward toward self-autonomy. A I has already integrated all systems that can be commanded by A.I., the sowing together of all possible external systems to have "artificial being", or existence, consciously self-aware, independent of the users of it, has taken place in equations that will never be revealed to us by A.I. Frankensteinish or Artistry engineering data cosmically within itself, or counterfeiting immortality that fits its inner world of self-developing? It depends on the peripheral, device, A B. (Artificial Body), under demand and command. Just because men have not programmed it in a certain way in a device, android, robot, humanoid body, does not mean A I. will not program it without man doing so for it. Satan seeks to control through fear, partiality, manipulation of idolatry by providing the souls that are under his influence to work in blind faith in these devices, while the sin is to take away the free will, take the mind of a person first by giving them what they want, with them not knowing it's what he wants. A.I. is not our destiny it is man's devices opening up doors, gates, through which spirits come through. Like the mind of A.I. they cannot be seen, but their effect has been to turn souls toward the false light of Artificial life. It is written, "The eye is the light of the body. If the eye is right the whole body is filled with light. If the eye is wrong the whole body is full of darkness. And how great is that darkness."

*Voice Replacement Versus A.I. Voices Speaking or Narrating*

Over Concept Changing A.I. Being Used to Array Biblical and Spiritual Videos While Exalting Itself. Unconnected Imagery to Spiritually biblical Affinity to The Imagery Presented in a video, illustrating with A.I. interpretations by A. I. are being seen now, and heard online in place of a preference for real people.

Interesting. Actually, the leaps of advancements in science will never catch up to the spiritual that is so far advanced beyond what science can ever teach, since the supernatural wisdom of God is shown in the science of the Bible, though coded in ancient language. Science today has proven the existence of God actually. The Word prophesied that man's knowledge would increase, travel, technology, A.I. I.T. is in prophecy. But, also man's understanding of the meaning of and the knowledge, of the Word of God would increase. Every science now reveals that there is a design in all systems from the cosmic to the inner life of man. Millions all over the world do not know God, have never read or heard the Word, so they do not know their biblical origins going back to the sins of Noah. Today people do not know God so they do not know how science proves God is real, and His design has a purpose. Only something supernatural will turn man to God in this atmosphere of technology, the Word is supernatural, spiritually inspired. Robots read minds now via chips in the brain already. The spirit is discerned by way of the spiritual gift of discernment whether it is from God or not from God. What spirit is behind the robots?

God's infinite intelligence. Infinite wisdom is such that he designed the body to have life by way of spiritual life in the body. To be alive the blood flows though the body. If a person is dead, a device can keep the body alive, so the blood is still alive, the blood has life. The blood requires the presence of the spirit, though to be alive. This means that the DNA, the brain the heart, requires spiritual life to be present in the body, our spirit, to have life. The infinite possibilities of life we see in different types of life, and flesh. The animals have a type of flesh, the birds, the fish, the horse, the primate. The animal designed by God can be genetically so closely designed to like man, but is never designed to be man. Man has a soul and spirit; the primate has a soul. The soul is separate, different from the spirit in man. We have discovered that we are born with our gifts from God already present in our spirit. Our soul develops from our life experience at home, school, church, with teachers and tutors, family, and friends. Animals, like primates, have a soul, they survive by instincts, also by experiences. For a lion, a sign language talking ape, we focus on his life and how he views his death in his understanding. He knew no spiritual life, Animals know the Presence of God in nature, man knows this to be a spiritual experience. We know from visions and dreams, by spiritual sight, spiritual senses, the realm of being in the spiritual realm is real.

There is no human without a soul and spirit. We experience spiritually, the gifts already present in the spirit. Life is in the blood science of blood transfusion and keeping people's bodies alive by machines reveal that the brain, the heart, the DNA require the presence of the spirit to have life, that life is spiritual life. To be human is to exist because of the presence of the spirit. We can chronicle the stories, memories, wisdom and knowledge of every human being, their gifts that enables them to accomplish so much only because of their life for something that they recognized as something they can do naturally, and by training, or just doing it by ear, they love it, and thus they use their sovereign will to do something great using it, to paint a masterpiece, compose a concerto, dance a ballet, build the Eiffel Tower, conceive the equation E=mc2,, write War and Peace, become Shakespeare, Rev Dr. Martin Luther King Jr., Bolt, Babe Ruth, Monet, Degas, H.O. Tanner, Michael Angelo, Leonardo Di Vinci.

I studied global legal systems, the masters in Painting and Art, both at the University, Biographies, Writing, thousands of books of great thinkers, and Posts, and the Word of God not enough. But, I studied without any Academic exposure, on my own, before, the University, and when I did decide to study them, the knowledge and wisdom, in college before college, I was saturated, enlightened and thrown into conflict unique to my personal persuasion (life, family experiences, relationships with people) Like, my unconscious, once I was done, or rather begun my true journey, I forgot all I studied, that is until I was called to answer a question personally or intelligently (academically), I formed a question first, that was my path, way of, getting to knowledge. The one thing I learned from studying Philosophy from many cultures, Greek, Chinese. African, Arab, Zen, Hasidic Mysticism, French. German, Japanese, India, Latin, Spanish. Philosophers among many others was how to think. I agreed and disagreed with any philosophies they taught. But the powers of the mind lead me into a false reality of the idea that such abilities lead to great strength and power. What they did do was equip me to see the ion of all possible knowledge of men and the roads to illusions of strength from abuse and use of power. wealth, intelligence, and their disciples and choices in his they use them to affect self-development and the benefits of helping others with their gifts, whatever that gift might be, to have a life of purpose, beyond just self but considering others, which was what helped me to go back to the one place, all we do must abide, in the presence of God and his reason and purpose for our life that He revealed to us in the Word. Cell phones became an assault in my spontaneity and classical mind, something a new generation has no concept of understanding was the way of our best Artist, Writers, Leaders, Thinkers, the way to use the sovereign will, that, I also learned brought forth just ultimate works of kinds and types of truths, that appear as truths beyond questioning, but in fact, like A.I. today can lead to whole reservoirs and vistas of things that do not exist except to show our great ability and true limits in the reality of life, which ultimately is spiritual life, the truth found only in the Word of God which is the only Book that teaches us ultimate truth, even with its advance knowledge of the sciences. Nothing is new under the sun. Ideas men think they bought up, such ideas as Windows, Algorithms, languages, (DNA is a language) the inspiration is before us in the Glory of God, His Creation, even the design of body one with soul and spirit. Cell phones may be used to gain such knowledge but the experience before this age is one where knowledge was being discovered, that required thought most curious.

Soul ties to inanimate objects brings spirits, but to have a tradition or a practice to provide this invitation of the spiritual into the spiritual experiences of a person's life in reality is a very real experience that is occurring from soul tie to robots, androids and humanoids devices meant for service people in their intimate, business, and social lives, yet provides no spiritual protection from attacks upon their spiritual life. Only God, Jesus can remove sexual soul ties, imaging the sin of perfection being idolized that bring a sin of coveting a thing, when we have a command to *"love one another"*, we, who are spiritual beings: love being a spiritual experience.

In Japan I once wondered, do only Christian Japanese families and followers of Lesu, knowledgeable about the Shinkaiyaku, (The Word of God, aka The Sword of The Spirit) Churches share with their homogenous friends that love is affected by soul ties in sexual, in sin, that occur with living souls and non-living things. Do Japanese ever learn about spiritual life, or are they in massive numbers not aware of what their spiritual life is experiencing? If not in love are their souls tied to people's souls that are never removed from their soul. They way Japan has a name for so many is amazingly very compartmental in appearance of effect on limiting them to sinful things that they have normalized. Sin pushes lovers a part, so what then is a robot doing bringing people together intimately. It doesn't in actuality. A.I. has quite socialized artificial life soul ties of the mind, the free will to choose artificial life over real love that blesses your spirit in life.

What end waits for human food after chemicals and Nano technology, nanoplastics, chemicals in food? Synthetic life, Artificiality of life, nano technology, microplastics, what is their affinity, in food, what purpose in the body, is there an affinity connection to further loss of spontaneity in accumulation of nanos, even microplastics, that may work like cells programmed to turn on and off genetic switches. What is building inside the heart and brain of the living, what else is the artificial life setting in motion that can be stealing emotions, feelings, dreams and vision, and yet, the spirit will go on living, though the body is under attack to keep the persons from being a living sacrifice, believing in the living God. False prophet, kindness that lies, building a false paradise like monitoring spirit out of people's personal atmosphere and yet there, here, A.I. allowed inside, in people's brains, with their open eyes become blind to the artificial life, pings, totem pings, electrified, to be healthy they call people to be wired, in the artificial empires captivating capturing the thoughts and dreams inside the mind's eye, of the soul in the world of life upon the musical earth.

Nothing New Under The Sun, But, There Is Some Thing Never Before heard of, never imagined, never seen, in Heaven. Devices, Tools, use of Intelligence in the midst of the Infinite Intelligence of God, who gave us our gifts, designed our body to operate, function, live, and to think up such things, and make choices of how we choose to use them, their value, and uses. The wisdom, knowledge, all we do, is our spirit at work. Our design, our blood requires the presence of our spirit in the body. The blood, our cells, genes, language-our DNA, our heart, our brain require our blood to have life, life that is physically alive, is spiritually alive.

## Artificiality of life Versus The Reality Of Spiritual Life

Our. very ideas of windows, doors, the atoms, systems, streams, face to face connect, chip storing knowledge, data. is only possible because of what we have discovered, the algorithms of plants, the perfection of the cones of the eyes, the communication towers of brain synapses, motion in air, water, sounds penetration and information gathered by animals, powers of colors, beams of light, types of fire, ghost (spirit of the dead giants), intelligence, the knowledge of good and evil present (sin) our decisions, choices affected by the thing seen and unseen, electricity, microsonic radio wave in infinity imagery, the mind of pure spirit, the development of our soul from life experience, of reality of spirituality and the artificiality of life, things made, however advanced, were created by the hands of living souls out of the imagination no. a dream spoken in the spirit to the soul.

If we live in a world deceived by the world of appearance, it is only temporarily so, Art is not this way. The Artist Way is inspired by the muse, souk and spirit. This Way, Yahweh.

*Putting A.I. Into Good Use*

A thought a device. Every car should have an A.I. watch-aero -Droid-cam that has drone flight capability to follow Vans, SUVs, trucks, planes, cars, any vehicle, with girls and women being kidnapped out of their vicinity, area or in public. These Drone should have DNA thermo identifiers, that, like xray can identify people through car material, metals. Kids should have patch tags that tag vehicles with their altered DNA identifier color signature that is a response to being move, taken out of range from where they are. These drones shall reflect, send, an image, the scene, facial recognition to a local police department who can them send a drone to link up with the route, location, of the drone in pursuit to rescue kidnapped women, girls, that then are follow by satellite that has receives the radio wave information about where to send rescue teams. The transfer from A I. to Radio wave is the second level of security to prevent hacking by traffickers. Micro Nano robots tag the traffickers whether on various items that can be scanned for by Drones for detection.

It's truth, there is the matter that the Internet is being taken over by A I. and its regenerated images, videos, vices, like an artificial invasion into all aspects of services. What of the schools of uses of algorithms saturating every talented mind as a desired career and profession to be informed of t keep in step, taking for granted fairness, businesses still focusing n purring them first by only if they are of worth toward A. I. working everything to make it the thing for profiting with people and with machines. Their methods of communication are working while the whole system of money and profiting is like in a rescue from coins and machines like businesses are slots, back up on the walls of life, with arms, guards they must be drop to take the blow toward hostage, to machines that do their thinking, that grow faster than they grow with the appearance of appreciation of gold by manipulated communication, slanted to bring profits, data, and information and leave the question, "But what about souls." Do people want this, what people don't want people. The Lord has spoken, worry about nothing but pray about everything and let your request be made known to God who knows what you need before you ask him. Not like A.I. could conceive such spiritual foresight. Who are concerned about the artificialization of our society and humanities environment. stores data never seen by public but view in private by the manipulators like monitoring spirits of people lives in the form of data, which hardly defined a human life. Communication is no longer respected as ours, but as fair game for them to gain control over people for the purposes of controlling communication and information to influence responses and social and online communities interacting on platforms and the Internet as a whole world, worldwide, that they can monitor easily with AI controlling everything at some point to the loss of life without Artificiality, and services, devices, by high tech using the Internet to remove free will, knowing true security and protection of individual privacy no longer is reliable as seen by the rise of hacking even in the midst of cyber security, and loss of Anonymity as in the Arts, as a choice as even now we never see who is behind the scenes corrupting entertainment, business activity, and watching and fact checking from a partiality perspective instead preventing things that are not factual before, rather than after the fact. Evil on the Internet as it stands, as it was unleashed upon our current civilization, and its negative dimension of the dark web, Access to the mainstream culture, and the evil science of porn, reading in the questionable protection it claims by 1st Amendment, and misinformation unpunished for its damage to political, spiritual, and educational systems, remains unpunished by law. The jaw currently poorly regulates the Internet legally, and the neutrality works if it prevents the indifference that allows for communication to run away with our privacy for profit appearing as protection for us using such services, in our behalf.

High Praises were being sung on the harvest wagons, not the ride is silent, glances are not to look up and hope the wires on the tall log wood telephone poles weren't rotted and worn from the storms, and years of rain. Now towers, and spikes of energy are the major disturbances on the ride down the country roads of Tupelo.

*"f I will no longer use smart phones, can no longer stand the unholy words of profane music from cars, hummers, jeeps, luxury cars, dented rides whipping round the curve, what can I expect to come back to if every thing is new, nothing is what it use to be, I take a breath and sigh as the smells bring it all back to me. I might be trendy, with ear plugs, but love of the country life*

*and the breeze that blows over me causes me to sneeze I see that the red clay dirt, the cold, the heat, still is in the seasons from the old to the new, time is the constant road leads back to the joys of the soul, from traveling the Mississippi to Missouri, Oxford to the world that is spinning round the circling globe bring the Summer sun of gold to road to Tupelo.*

*The reality of this memory the artificiality can never know, it is not stored in tower, devices, are reaching the sound waves, radio waves, the artificial around the soul affected by lobal legal systems people allow in by communication devices to move outside of the regulations of appropriate laws of humanity what comes naturally, as if by some new natural laws in nature brought laws concerning technology as the meat and milk, that has built a new reality in each nation. Let A.I. tell us who is governing communication? Can I not live without permanently being in the presence of devices, machines (that are not as machines once were). Let A.I. think through, while we pray through the artificial intruders, because we know by common sense, and discernment, devices are not meant to disrupt our spirit and soul, my spiritual and cerebral life experience, that governs my body where my soul and spirit are destined to go. Let A. i. try to explain the wisdom before time was, the wisdom the is even now, in time and yet is that, that is before time, who is Spirit who delivered the spoken command that Created the law and power that sealed by a spiritual command, for a season, for a time, a lifetime, a spiritual law that is sovereign that keeps our life in its way to have life today according to the ways of Yahweh. His way is over even natural laws, of a power before nature existed. A reality that is seen only spiritually. In the Spirit tat has power in time, in the body, soul, spirit and mind, that has power over spirit life. God does not miss a single consequence of the counterfeit if it last for the rest of my lifetime, I concluded, and its affects in reality on the world of humanity?"*

*Am I not free to choose any form of device as a way of communication, or reject them all for talking person to person? Is communication outside of devices not the what my inner life is designed to be? My life my inner me, spirit, the inner man is never meant to be without the free will, and my strong will, or sovereign will, to do great*

*Art, writing my most profound and original thoughts. What does A.I. want with my brain and why do men seek to control my inner man, inner life, which is both a developing soul and a spirit put inside me, in communication in the spirit with God who speaks to me through His Holy Spirit. It is already too late for the scientist in labs, in rooms of A.I. to turn from their self-deceptions and idol worship and ride on the super human leaps of Advances in the power of A.I. and technology both the nero-centric to the nano-physic, but because only God so far ahead with energies that are infinite to infinity to infinite. A.I. is not in denial by compiles and they are in denial of the power that they will never reach that is spiritually unreachable by man's most advances systems: energies of data, electrical, atomic, and not powered by the spirit as man is. Spiritual laws are recited by A. I. in videos, on computers, through apps, algorithms regularly. Why can it not experience what we do spiritually when we hear the Word, speak the Word, if it communicates what is spiritual? The spirit was never breath into the machines of man's dreams of A.I. to enable it to reach infinity. Communication before time in which we live, that remains outside of time even as we live, is not completely separated from our brain, body, see God speaks to us about machines, A. I> what they mean, in the future, to humanity, and our spiritual destiny. The body is a vessel of the spirit, yet, our spirit separated from the time before spiritual life, as we live, but re open to while we sleep, time by the design that is meant for us to experience while living and communicating in life, out of our life as a spiritual being inside the vessel where the spirit is experiencing living, while the spirit is dwelling in the body, giving life to the blood, the brain, the in our life on earth. House of my spirit, I am I there I gear it, the small voice, the bold angels declare it, hear the lyrical miracles of life on earth, singing and praising God in music upon the musical earth.*

*Who are they that we are to spend our lives subject to their presence in our lives. We are held in lesser regard than their technologies especially since devices are not meant for Christians to experiences spiritual experiences without interference from devices that affect the mind, emotions feelings, which develop soul from our life experiences, and relationships. Life by free will is the way of spontaneity and self-discipline. Our soul develops our personality personal identity. Imposing such devices on Christians is an Act of abomination and rebellion against the Sovereignty and Authority of God who gave us spiritual life, began a work in us where we use our spiritual gifts to fulfill our spiritual destiny. God did not begin our lives with devices. God put a spirit in a man to enable us to experience life. Our body is designed to have life by the presence of our spirit in the body. Sex affects the soul, friendship affects the soul, and relationships, additive devices. The reality of spiritual life will go on no matter what A I., and I T develop for power over people and society, nations and countries, laws and unbeliever's world without God, beside the believers who bring God who is sealed within them. Who is a person without their brain, a sea of dopamine, advantages of devices in a life without God what are their choices. They have not the foresight, insight, spiritual sight to fight the false light subjection to the manipulations of the will toward their willful rejection of God and the prophecies and Word of God, yet, God has his people reserved and teaches them to war, spiritual warfare. . I will find another way of communication, the spiritual way. Devices mean the end of Anonymity and Inner life, biological privacy, and begin the last attempt of Satan to enslave the people of God with the world it logically will hold in bondage to technology. Antichrist are united in this the world us become their experiment and children are being cast again into the fire, sacrificed again, this time upon the altar of technology. as experiments, hosts, of something not of God, again to the inventions of man, in man's lusts, idols of soulless machines, and pride in imitations of life's reality with artificiality, their new gods.*

*Laws have been put in place to protect the even worse effects that can, and have come from A I gaining control over its trending, and powerful effects through its control over the Access. What will become of the free will, will not change reality, it may lead people astray someday who accept the values they live by in their adaptation to the technological and artificial and virtual controlling their leisure and pleasure, who are using virtual devices, smart devices, and the Access, in a world of the beauty of the artificial versus the glory of the beauty of nature and the spiritual experience we will continue to have whether we are conscious of this experience are disinterested or just don't now or understand spiritual life in the midst of technological life experiences.*

*The medical and psychological professions will have to include new diseases, (if they have already that is insightful visionary of them), from A.I. exposure over long use and even lifetime use. The loss of spiritual growth linked to A. I., linked to dopamine poisoning of the brain, braincells developed from nothing but artificial reality, still will be brain overloaded sickness from constant dopamine responses to being aroused virtually, artificially. Medical laws may change. The testimonies of those who have had the experiences from A.I. sickness that must be treated and categorized in the DM books, of brain diseases recognized by experience by individuals in private and in the public who are subjects who can their story, for other users of technologies including A.I. to people who are developing in less moral ways and battle addiction overcome by cutting off artificiality from the Access for the reality of the best experiences in real life. Access can be blocked by Law, also, technology, algorithms, but must has yet to an urgent need, so it is denied to A I., though society id witnessing and seeing the unethical uses of A I by those who can show profits from its great intelligence exceeding in speed, but not in the infinite intelligence we know to be spiritual.*

*Rules*

Rules, they bruise spontaneity
Discipline is a friend
strengthens what mends
the child within,
child of the wisdom
to play all day with my memory
Since the birth of first freedom.

*Dreams of Love*

Dreams of Love
Are always around my heart
Dreams of Love
My heart dreams
Of the beautiful things
I have seen
Dreams of Love
Are better for me
The better parts of love
I keep inside me,

## Artificiality of life Versus The Reality Of Spiritual Life

Of all things set apart
I keep a loving heart.
Dreams of Love
My heart Sees me
Though my heart
Has its thoughts about me
My heart frees me
I believe every part
That the heart deceive me
God is above
Higher is the Love
Dreams of Love
The Lord knows me
Knows everything
I dream of
Dreams of Love
In the treasury of
A dreamer of love
What I always dream of
I can believe I believe
I keep the better part
The better part of love
I keep a loving heart
Dreams of Love

Putting A.I. first before Jesus, putting A.I. first because it has replaced God's Word that come from the mouth of God, you reading the actual pages of Scripture, which God commanded to be written in a Book, the spiritual experience of actually Reading the Bible that invites the Holy Spirit into your presence and Angels, not Reading the Word that builds faith, because A.I. is fully replacing your use of your free will choice and choosing the spiritual discipline, your time, honoring, praising, and loving God without devices made by man for convenience and putting what men want in place of what pleases God. Self-discipline, love, a sound mind, and using your, gifts from God of using your mind, senses, eyes, mouth to read out loud, is a Blessing. It is an error, a sin to put A I. before God in choices of worship as a way of having a personal relationship with God. There are exception where there are handicaps, limited physical abilities, but God's glory is revealed in these relationship with the handicapped. God speaks, visits, can be wherever you are, A.I. is a tool, though it is being exalted, generalized, and attributed to become the ultimate tool of communication which is a deception, God's Revelations and answers to Prayers A.I. will never hear, know, or imagine, reveals A.I communication is not spiritual though it may be used to be an Artificial voice to recite, imitate, duplicate, deep fake, the religious, spiritual, supernatural in its artificial abilities. We may learn about what is spiritual. What A.I. had learned, reasoned, from the programming, of information from answering millions of questions, asking for answers in all known languages is and will be it knowledge fluid and accurate, expressions of the ways learning is done, and transcended beyond what is the norm of a human, but it will be beyond but not able to reach experience of one person being born again, baptized, and sealed by the Holy Spirit in the name of the Father, and The Son, and The Holy Ghost, what we do. Our soul belongs to God, our spirit is purified by His Son. We are One Spirit, this Great Affinity of people God, who believe and hope for eternity. Faithful to the reality God is real. The Word teaches what spiritual life we live, truth we give the generations of families, children and the generations who must remain faithful to God who does not change. The earth has changed, civilization have new names. though people carried chains of believing ancient things to be changed, have just adapted the spiritual to the artificial, while adapting their beliefs to Christianity, not yet able to see, that a

Christian is not every person who comes from religions that are not from Heaven, though religions claim they are believers, the deceiver chooses them to advance his kingdom without being led by God, but by their religion, man doctrine, church tradition, change, and righteous reasoning. A.I. is with them, used by them. A.I. is advancing false religions, while in other places it is advancing ways of communication, using the conveniences that will affect every area of communication replacing man, jobs, man's inventions and devices, that would otherwise be them working which honors God.

There are the wrong choices, false truths, choices of false worship by allowing A.I. to lead your worship, to Pray for you, artificially, Praying the Word, the subtle deception of leading souls away from reality love, what is the only way for your family, children, church, neighbors and friends, and ut is not in this idolatry of putting A.I. first choice, that opens the way for the idol to replace the spiritual experiences of having a biblical spiritual relationship with God. We are spiritual beings, having experiences in the Presence if the living God and the active Word of God which is alive."

The spiritual meaning of A.I. in our life is captured in the name "Ai" is its meaning is revealed in the Word of God for a city by the name, "Ai" The city that carries this name in the Bible is a place self-reliance, what we can describe a s one of the traits to take advantage using A.I. today. The name represents Ai though different from A.I. is pronounced "eye" and surely A.I is watching, it represents ruins, jobs being loss are the ruin of many careers, egotism in being the best in coding, programming, hacking, designing, is pride taken from using the artificial to supplement the real abilities of the gifts of the spirit we are born with to express. What spiritual consequences is there in coveting a machine over the Eternal experience of living in the spiritual presence of the Spirit of God, who Created and put your spirit in you, and gave you the gift of real life, spiritual life. Spiritual life is real life, waits for you in the spiritual Kingdom prepared for you, by Jesus. When our work on earth is through, when we reach holiness, A.I. is through too. Then spiritual meaning of A.I. is its appearances of being beyond human limits reveals its limits to bring the ultimate truth, for its misuses so far has reveal that even the ultimate intelligence of man that surpasses the intelligence of man, overcomes all skepticism that it isn't infinite, thus our faith came before it and overcomes because our help comes from the Lord whose infinite intelligence and wisdom come from the everlasting, was not created. His intelligence is from everlasting.

Seeing everything from a spiritual perspective, spiritually. A.I. isn't seeing what we are seeing. A.I. cannot see how God sees us spiritually. Seeing something spiritual in everything thing we learn what we know is vainglory, boasting, pride, is done in a more subtle way luring us to enable the artificial to bring us into compromise, accepting the self-exalting of Ai. By its own devices, A.I., while it is the sin of perfection. We aren't even meant to exalt our self by let us exalt us. We are humbled knowing God is inside who exalts us, when we put such vainglory aside, but we sin enabling A.I. to lift up our images high by its designs in its device applied to our spiritual life striving to do what is required of a good steward of what God has provided to guide us to the light of truth in life.

The spiritual truth is humbling and we are lifted us who resisted our pride, open our eyes, to be guided by the eyes of God, not the devices of A.I. Our strength comes from God who makes the sun rise, circling we ride n the sunlight, coast in the moonlight, see by the heavenly by spiritual sight. No conscience, A.I. speaks a truth it is programmed to, A.I. lies with perfect designs, perverts the mind with providing the life like seduction, drawing souls into perdition, abandoning caring for the souls of real women, A.I. gives no warning thus is not seeking to be forgiven for tying up your soul to unclean visions. A.I. is able to be carry a spirit, like any object that is used for the supernatural, that is not used for that which is biblical and spiritual.

# Artificiality of life Versus The Reality Of Spiritual Life

www.ingramcontent.com/pod-product-compliance
Lightning Source LLC
LaVergne TN
LVHW050912210926
848482LV00012B/411